OFF THE NETWORK

ELECTRONIC MEDIATIONS

Katherine Hayles, Mark Poster, and Samuel Weber, Series Editors

(continued on page 194)

OFF THE NETWORK

Disrupting the Digital World

ULISES ALI MEJIAS

ELECTRONIC MEDIATIONS, VOLUME 41

University of Minnesota Press
Minneapolis
London

The University of Minnesota Press gratefully acknowledges financial assistance provided for the publication of this book by the Office of Academic Affairs at the State University of New York, College at Oswego.

A different version of chapter 2 appeared as "The Limits of Networks as Models for Organizing the Social," *New Media and Society* 12, no. 4 (June 2010): 603–17. Portions of chapters 6 and 8 were previously published as "Liberation Technology and the Arab Spring: From Utopia to Atopia and Beyond," *Fibreculture* 20 (2012). A different version of chapter 7 appeared as "Peerless: The Ethics of P2P Network Disassembly" in the proceedings of the *4th Inclusiva.net Meeting: P2P Networks and Processes* (Madrid, July 2009).

Published by the University of Minnesota Press
111 Third Avenue South, Suite 290
Minneapolis, MN 55401–2520
http://www.upress.umn.edu

ISBN 978-0-8166-7899-0 (hc)
ISBN 978-0-8166-7900-3 (pb)

A Cataloging-in-Publication record for this book is available from the Library of Congress.

Printed in the United States of America on acid-free paper

The University of Minnesota is an equal-opportunity educator and employer.

20 19 18 17 16 15 14 13   10 9 8 7 6 5 4 3 2 1

FOR ASMA

CONTENTS

ACKNOWLEDGMENTS

Gilles Deleuze observed, "We write only at the frontiers of our knowledge, at the border which separates our knowledge from our ignorance and transforms the one into the other."[1] This attempt to breach the frontiers of my own ignorance was aided by the Grace of the Guide but also by many people whose support and knowledge greatly contributed to the work.

First, I want to thank the editorial team at the University of Minnesota Press, in particular Doug Armato and Danielle Kasprzak. Their efforts helped to make this, my first book publication, a gratifying experience. I am also very grateful for the support and encouragement I've received from people at SUNY Oswego, including Dean Fritz Messere, Provost Lorrie Clemo, and the many wonderful colleagues and friends from across the college who have inspired and motivated me.

The point of departure for this project was a set of queries originally posed in my dissertation, "Networked Proximity: ICTs and the Mediation of Nearness," undertaken at Teachers College, Columbia University, under the guidance of Robbie McClintock, Hugh Cline, and Frank Moretti. Their tutelage during that initial stage is much appreciated. Some of the initial ideas for this work also emerged during a 2008 Visiting Research Fellowship at the University of Amsterdam Business School, Program for Research in Information Management. I am thankful to Rik Maes for facilitating the visit and being such a good host.

The book benefitted from the insights, critiques, and corrections of many who devoted their time to reading early drafts or providing comments during conferences. My spring 2011 Social Networks and the Web class at SUNY Oswego reviewed the first two chapters, and I thank them for being both critical and enthusiastic. I acknowledge Tiziana Terranova and Nick Couldry for comments offered during the 2011 Platform Politics conference in Cambridge, and Michel Bauwens and Juan Martín Prada for their reactions to a presentation made in 2009 at the fourth Inclusivanet meeting in Madrid. Electronic correspondence with Trebor Scholz forced me to clarify my position and improve the framework for my argument. Zillah Eisenstein and Geert Lovink read the manuscript draft and provided valuable feedback, and they were extremely generous mentors as well. The two reviewers assigned by the University of Minnesota Press, Nick Dyer-Witheford and Jodi Dean, contributed detailed responses that helped to strengthen my argument. Thanks also to Anna Reading, Hart Cohen, and especially Ned Rossiter for inviting me to the University of Western Sydney to share my work. The conversations I had with everyone I met during my visit were encouraging and instructive. Without the help of all these people, this project would not have been possible. I am grateful for their assistance.

Additionally, as I learned, one cannot write without a strong support network, and I was fortunate to have around me friends and family who never failed to inspire and encourage me. It would be impossible to list all of them here. But I cannot fail to thank Madhavi Menon and Gil Harris, whose friendship is a gift and who continuously provided encouragement and good advice (thanks also to Madhavi for reading parts of the draft and making helpful suggestions).

I also want to specially thank my parents, Elizabeth and Manuel, for always believing in me and doing their best to nourish my intellectual curiosity. There are no words or deeds that would suffice to express my gratitude and love. I should add that this book is very much written with my cherished nieces and grandchildren (the younger, networked generation) in mind: Abril, Ilse, Ana Elena, Mina, and Batu.

Most important, I thank my beloved wife, Asma. As my "resident critic," she forced me to develop ideas and improve arguments. As my "in-house editor," she painstakingly reviewed the entire manuscript and offered suggestions on everything from the writing style to the structural aspects of the argument. As my private mentor, she provided advice and support during the more difficult stages of the writing and publishing process. As my loving partner and companion, she was truly a source of inspiration, comfort, and well-being. She nourishes not only my mind but also my heart, and because of this I dedicate this work to her.

INTRODUCTION

> Networks matter because they are the underlying structure of our lives. And without understanding their logic we cannot change their programmes to harness their flexibility to our hopes, instead of relentlessly adapting ourselves to the instructions received from their unseen codes. Networks are the Matrix.
>
> **MANUEL CASTELLS,** "WHY NETWORKS MATTER"

ON MAY 31, 2010, an estimated thirty-three thousand people[1] committed suicide in a collective wave of global proportions. In the opinion of the media, however, the aggregated death of those thousands was essentially insignificant.[2] Thankfully, no blood was spilled that day, since the act of annihilation in question involved permanently deleting one's Facebook account in what came to be known as Quit Facebook Day— an expression of rage over the company's privacy policies for some, and of disillusionment with virtual life for others. In the words of an early advocate, "The movement could reach epidemic levels if more users kill off their electronic selves rather than submit to corporate control over their friendships. Facebook, and the other corporate lackeys, will then learn that they can't exploit our social relationships for profit. From viral growth will come a viral death as more people demand that Facebook dies so our friendships may thrive."[3]

Availing themselves of how-to advice from the movement's main website (Quitfacebookday.com), as well as tools like the Web 2.0 Suicide Machine (Suicidemachine.org), people removed themselves from the popular social networking site because they agreed with the general sentiment that "Facebook doesn't respect you, your personal data, or the future of the web."[4]

While thirty-three thousand is a trivial portion of what was then a five hundred million membership base, Quit Facebook Day was deemed a success even as it failed. The mass exodus that was hoped for did not materialize, but at least the movement generated a public relations disturbance that led Facebook to reconsider its policies or at least to try to do a better job of explaining them. Thus the events surrounding Quit Facebook Day shed some light on today's frequently tense relation between the rights of the user and the interests of the corporations that operate digital social networks.

Quit Facebook Day, as an expression of the desire to kill one's networked self, illustrates the need for a language to talk about these tensions, to talk about the darker aspects of the relationship between platforms and individuals. It is obvious that digital information and communication technologies, such as Facebook, act as templates for organizing sociality, for building social networks. They arrange individuals into social structures, actively shaping how they interact with the world. But during the process of assembling a community, not every type of participant or every kind of participation is supported by the technology. While some things can be assimilated or rendered in terms that can be understood by the network, others cannot. As participation in social and civic life becomes increasingly mediated by digital networks, we are confronted by a series of disquieting questions: What does the digital network include in the process of forming an assemblage and, more important, what does it leave out? How does the network's logic of exclusion shape the way we look at the world? At what point does the exclusion carried out by the digital network make it necessary to question its logic and even dismantle it, and to what end exactly? These are the questions this book seeks to address.

A network, defined minimally, is a system of linked elements or nodes. While a network can be used to describe and study natural as well as social phenomena (everything from cells to transnational corporations), what is relevant here is the use of networks to describe—and give shape to—social systems linked by digital technologies. For our present purposes, then, any and all kinds of electronic technosocial systems will simply be referred to as "the digital network." We can broadly define a digital network as a composite of human and technological actors (the nodes) linked together by social and physical ties (the links) that allow for the transfer of information among some or all of these actors.[5] While the Internet is the most notorious example of a digital network—and the main focus of attention in this book—digital networks can encompass other technologies not based on the Internet, technologies such as mobile phones, radio-frequency identification (RFID) devices, and so on. To make this analysis as broadly applicable as possible, however, the collective label of "digital network" will be used to encompass both the Internet and other assemblages constituted by various digital information and communication technologies.

While not unproblematic, the conceptual grouping of all digital networks into a discussion of *the* network is, I believe, timely and necessary. Modern contributions to social theory, science and technology studies, and even critical theory[6] have shown us that networks are plural, fluid, and overlapping; we do not belong to a single network, but to a variety of them, and our participation in them is variegated and complex. To propose a critique of *the* digital network might seem, therefore, to reify, essentialize, and reduce the object being questioned. But as I will be arguing throughout this book, it has become necessary to isolate the network as a single epistemic form in order to launch a comprehensive critique of it. We have indeed gained a lot by looking at the world as a plurality of networks. But we are starting to lose something in terms of identifying common characteristics and, more important, common forms of violence found across all forms of networked participation. The essentialism behind discussing the network, therefore, is a strategy meant to clarify the relationship between capitalism and the architecture of digital networks across a variety of instances; to facilitate, in short, a structural critique or *unmapping* of the network.

Why talk about unmapping the digital network in the first place? The very project that the title of this book suggests seems unnecessarily antagonistic at a time when it is almost universally accepted that digital networks—everything from cell phones to social networking sites—are bringing humanity closer. At least this would appear to be the case if we go merely by adoption rates. More than a quarter of the world's 6.7 billion people are already using the Internet.[7] With only a few exceptions, Internet penetration has surpassed 50 percent of the population in most of the thirty countries that belong to the Organization for Economic Cooperation and Development.[8] And while developing nations obviously continue to face a digital divide (e.g., there are 246 million Internet users in North America, while only 137 million in Latin America[9]), they are by no means unconnected: according to a UN report, there are 4.1 billion mobile phone subscribers worldwide,[10] which means more than half of the planet's population now owns a cell phone; in Africa alone, 90 percent of all telephone services are now provided by mobile phones.[11] In the face of all this connectivity, any talk about *undoing* digital networks—however theoretical it might be—seems to suggest a halt to this march of progress.

Furthermore, critiquing the digital network would seem like critiquing the creativity and entrepreneurial spirit of the corporations that brought us the information revolution. If anything, the media seems to be telling us that this should be a time to celebrate and emulate the success of these digital captains of industry: Google, incorporated in 1998, now has a market value of $200 billion; Facebook, launched in 2004, now has the biggest social networking service, with more than a billion users, growing by 5 percent a month. There are social media pioneers like Twitter and Tumblr that have redefined the way we communicate, hardware companies like Apple and Cisco that have redesigned the devices needed to access the network, and even "old guard" telecom companies like Comcast and Time Warner that make it possible for us to connect to the wired world. These companies are economic forces, industry innovators, and, some would say, cultural icons. Our lifestyles (and in many cases, our livelihoods) depend on them. Yes, increased competition in the marketplace and stronger consumer advocacy would be welcome, but there is no denying that the information revolution these companies have facilitated is changing the world.

To find supporting evidence for this sentiment, one need do nothing more than to take a quick look at recent titles in the computer and Internet culture section of any bookstore (which would probably be done online, anyway). The volumes suggest that, among other things, digital networks are revolutionizing the way commerce,[12] domestic and foreign politics,[13] socioeconomic development,[14] and education[15] work. In the midst of this wave of improvement, with networks seemingly making possible practical solutions to many of the major problems that we face, is it not irresponsible to question their power? Yet in the chapters to come I attempt to do just that, find the motivations and conditions under which it becomes not only desirable but also necessary to disidentify from the digital network. But why?

Jacques Ellul proposed that whereas "primitive man" was socially determined by taboos, rites, and rules, the technological phenomenon represents the most dangerous form of determinism in the modern age.[16] Our tools shape our ways of acting, knowing, and being in the world, but some of their influence can unfold without our consent or even awareness, and this determinism is particularly dangerous. Thus to Ellul technology occupies today the place rites and rules did before modernity, both because they direct our actions and because they frequently go unquestioned. Without even realizing it, we become slaves not so much to the technology, but to the assumptions about what they are for, what they do for us, and so on. The goal of this book, therefore, is to attempt to specify the kind of threat that the determinism of the digital network poses.

Organization of the Book

The book is divided into three main parts. The first part, "Thinking the Network" (chapters 1 through 4) concerns how networks shape us, and how we, in turn, shape them. Chapter 1 ("The Network as Method for Organizing the World") introduces the notion of the network as a template for knowing and acting up the world and establishes the initial framework for arguing that the logic of the network (with its nodocentric politics of inclusion and exclusion) is part of a capitalist order that exacerbates disparity. Chapter 2 ("The Privatization of Social Life") engages in an examination of the political economy of networks and the process of commodification that allows them to increase participation while simultaneously increasing inequality. Digital networks, it is argued, are not that different from other for-profit media systems in the patterns of ownership conglomeration they exhibit, insofar as these corporations strive to eliminate competition in order to acquire larger audiences. The chapter thus proposes that monopsony (a form of competition characterized by many sellers and one buyer) has emerged as the dominant market structure in the era of user-generated content. A critique of participatory culture is put forth that frames it as both a form of pleasure and a form of violence that subordinates the social to economic interests. Chapter 3 ("Computers as Socializing Tools") takes a closer look at the scientific and technological paradigms behind digital networks and how they have been applied in the assemblage of digital social networks. Since a true understanding of digital networks is impossible without a good grasp of modern network science, the scientific study of networks—with its discrete set of metrics and measures—is discussed as an exercise not just in describing social networks but in designing them. In chapter 4 ("Acting Inside and Outside the Network"), the relationship between the network and the self is considered in more detail. Specific biases in the manner in which the network mediates the social reality of the individual in terms of immediacy, intensity, intimacy, and simultaneity are discussed. Different models for conceptualizing how the network and the individual codetermine opportunities for action are reviewed, including actor–network theory. The chapter then looks at how the network shapes the individual's opportunities for political action. The question of whether digital networks promote the formation of publics or masses is addressed as a way to introduce a discussion of whether the network has come to replace or merely supplement the role of the state.

The second part of the book, "Unthinking the Network" (chapters 5, 6, and 7) begins to address the issue of how and why unthinking the network episteme is necessary and possible. Chapter 5 ("Strategies for Disrupting Networks") lays out the theoretical grounds for doing this by discussing an ontology that accounts for the virtuality of networks. Digital networks give shape to social forms that were before only virtual possibilities. However, in the process of actualizing them (giving them concrete form as templates), they become rigidified social behaviors. Using the work of Gilles Deleuze, the chapter explores how the process of unmapping the digital network involves reengaging the virtuality of possibilities. This chapter also theorizes some general tactics for unmapping the network (obstruction, interference, misinformation, intensification, etc.), identifies the analytical spaces where such strategies can be applied, and suggests the personal and collective stances that unmapping might entail. Chapter 6 ("Proximity and Conflict") begins to examine the motivations for unmapping the digital network by focusing on the concepts of space and surveillance. While the uniform distancelessness of nodocentric space does not diminish social opportunities, it changes what counts as proximal and relevant and redefines our relationship with the local, and therefore must be questioned. Similarly, the chapter considers how network logic has changed the way in which dissent, security, and war are manifested and countered, and asks what some of the implications of this new order are. Chapter 7 ("Collaboration and Freedom") applies a similar approach to unthinking the network episteme when it comes to discourses related to commons-based social production and Internet freedom. The chapter questions the efficacy of peer-to-peer as a mode of social production that attempts to democratize resources. This mode exemplifies the limits of applying network logic to unthink networks because it simply manages to build a digital commons on top of an infrastructure that is thoroughly privatized. Likewise, the contradictions in the trope of "Internet Freedom"—as exemplified in the speech made in early 2010 by Secretary of State Hillary R. Clinton—are carefully scrutinized. The capitalist state and the corporation are typically portrayed as the stewards of the Internet, in charge of guaranteeing the rights of global citizens to freedom of speech, economic opportunity, and so on. In practice, however, the chapter examines how their actions undermine the rights and autonomy of individuals by utilizing digital networks to promote surveillance, repression of minority voices, and disparities.

Of the strategies for unmapping the network, one that might be particularly productive is intensification, since it involves not rejecting the digital network but using its own logic to subvert it, in the process creating alternative models of subjectivity that change what it means to participate in the network. This is the approach that concerns the third and final part of the book, "Intensifying the Network." Chapter 8 ("The Limits of Liberation Technologies") discusses the use of digital networks during the Arab Spring movements to point out how certain discourses prevent a critique of the tools and the market structures in which they operate. In this chapter, I also review some experimental work I am doing with alternate reality games as educational tools for intensifying the digital network. Chapter 9 ("The Outside of Networks as a Method for Acting in the World") expands the discussion of intensification by focusing on the importance of the outsides of networks and offers a conclusion that provides additional thoughts about the unmapping of networked participation.

While this is a book about ideas and concepts, I have tried my best to stay away from the overly abstract language that often accompanies the formulation of critical theory. If, indeed, there is nothing more practical than a good theory, as Kurt Lewin suggests,[17] I have endeavored to make the ideas in this book as clear and applicable to as many different types of readers as possible.

I

THINKING THE NETWORK

If there is no longer a place that can be recognized as outside, we must be against in every place.

MICHAEL HARDT AND **ANTONIO NEGRI**, *EMPIRE*

How is an ethical and political act possible when there is no outside?

BÜLENT DIKEN AND **CARSTEN BAGGE LAUSTSEN**, "ENJOY YOUR FIGHT!: 'FIGHT CLUB' AS A SYMPTOM OF THE NETWORK SOCIETY"

1 THE NETWORK AS METHOD
FOR ORGANIZING THE WORLD

THIS BOOK INVESTIGATES how the digital network forms part of a capitalist order that reproduces inequality through participation and how this participation exhibits a hegemonic and consensual nature. It describes the emergence of a network episteme that organizes knowledge according to reductionist logic and exposes the limits of trying to counter this logic on its own terms. Additionally, it explores the motivations and strategies for "unmapping the network," a process of generating difference and disidentification. While these themes are considered in detail in subsequent chapters, here I will attempt to establish a general framework for their discussion.

The digital network is a particularly delusive technological determinant because it is a mechanism for disenfranchisement through involvement and for increasing voluntary social participation while simultaneously maintaining or deepening inequalities. In other words, while the digital network increases the means of participation in society—as celebrated in much of the current literature—it also increases socioeconomic inequality in ways that we have not yet fully begun to understand. Networks are designed to attract participation, but the more we participate in them, the more inequality and disparity they produce. The way in which they do so—the way in which they create inequality while increasing participation—is through strategies that include the commodification of social labor (bringing activities we used to perform outside the market into the market), the privatization of social spaces (eradicating public spaces and replacing them with "enhanced" private spaces), and the surveillance of dissenters (through new methods of data mining and monitoring). Various examples of these dynamics will be discussed throughout the book.

This is not to say that participation in digital networks fails to yield any benefits, for it does produce many gains for participants. For instance, participation may increase social capital, such as rank within a community, or attention capital,[1] such as the number of times one's profile in a social networking site is viewed—all of which explains why some nodes have managed to "make it big" with very few resources in what appears to be a level playing field. But my point is that these methods of capturing and measuring new kinds of social wealth are means of concealing the fact that participation in the network promotes, overall, a kind of inequality that can eventually nullify most of its benefits.

Inequality is, in fact, part of the natural order of networks, particularly those exhibiting a *preferential attachment* process. The outcome of this process—whether we are talking about networks of proteins, citations, or web links—is that the rich nodes in those networks tend to get richer. This is not something that should strike us as illogical or irrational, since we know that even (or especially) in the midst of great disparity, those with resources manage to increase their wealth at the expense of those with fewer resources (which explains why it was recently reported that the world's rich got richer amid the worst recession in decades[2]). What I am interested in, therefore, is looking at the natural and artificial properties of digital networks that generate inequality and exploring their social, political, and economic impact both within the network and beyond it. In other words, I am interested in a political economy of participation in digital networks: looking at how the act of participation in digital networks increases the wealth of the corporations that own the networks and fails to generate any substantial long-term gains for the participants, even though it might seem to generate some short-term gains.

The starting premise, as many authors who have written about the information society have argued, is that the network has become the dominant operating logic of late capitalism. Michael Hardt and Antonio Negri, for instance, write that "[i]n the passage to the informational economy, the assembly line has been replaced by the network as the organizational model of production, transforming the forms of cooperation and communication within each productive site and among productive sites."[3] But the network has become much more than a capitalist organizational paradigm. It has become the means through which capitalism (which produces inequality as a by-product of the generation of wealth) can profit from social exchange and cultural production. This is possible because the network facilitates what Mark Andrejevic calls a *digital enclosure*.[4] Much like the transition from feudalism to capitalism involved the appropriation or enclosure of communal lands by private interests, today's digital enclosure also commodifies the public—not in the form of land, but in the form of speech and social acts—and widens the economic gap between those who own the means of production (the digital networks) and "those who sell their labor for access to those means" (labor, in this context, means participation in the network, which generates user information that "becomes the property of private companies that can store, aggregate, sort, and in many cases, sell the information to others in the form of a database or a cybernetic commodity").[5]

Thus digital networks are oppressive not by virtue of being digital or being networks per se but by virtue of being part of a capitalist order that produces inequality. The unfairness and inequality of participation in digital networks is a difficult trend to observe given the fact that an increase in access to digital networks is, most of the time, reported as a sign of progress. In order to provide a clearer picture of this inequality, we must consider not only arguments that show the immediate benefits of a particular technology but also broader arguments that contrast the increase of access and participation with more comprehensive societal indicators. For instance, a Pew Internet and American Life Project survey from July 2010 indicated that cell phone ownership in the United States was higher among Latinos and African Americans (87 percent) than among whites (80 percent).[6] This would seem to suggest some progress in terms of inclusion and perhaps even economic opportunity. However, when we contrast these data with the fact that the median wealth of African Americans decreased 77 percent from 2007 to 2010 (in 2009, it was $2,200 compared to a median net worth for white households of $97,900[7]), it becomes apparent that access to the digital network does not, by itself, translate into more equality. It might thus be helpful to speak of the inequality generated through participation via digital networks in the manner that Andre Gunder Frank[8] spoke of underdevelopment: not as the result of being excluded from the economic systems of capitalism, but precisely as the result of being included and participating in them.

Participation in digital networks produces inequality because it is asymmetrical. For instance, while users surrender their privacy for the sake of convenience, network owners are increasingly opaque about the ways in which they use the information they collect, as Andrejevic suggests.[9] The full range of inequalities that participation in digital networks can produce has not been fully indexed, but it includes dynamics such as the transformation of public goods into private goods once they are uploaded to the network (think of the LOLCats.com model); the way in which small social media projects are acquired by corporations who capitalize on the social labor of the site's existing communities (like Yahoo! in the case of Delicious.com), in some cases only to later disband those communities when the parent company experiences financial hardship; the warrantless monitoring and surveillance of action and speech as users participate in networks; and so on. These and many other examples can be used to build a picture of the inequality networks are generating. But rather than proceed merely by documenting examples, my goal in this book is to build a theoretical framework for understanding how inequality is produced and, more important, how it can be disrupted. While it would be valuable to quantify how participation in digital networks makes people poorer, we must begin by theorizing how the digital network converts our participation into disparity in the first place.

One of the ways in which it does this is through the *commodification of the social*—that is, by delegating more and more social processes to the market. If certain social functions before were performed in the public sphere and they are facilitated by for-profit digital networks now, or if new social functions emerge that can only be facilitated by for-profit digital networks, it means those social functions have been commodified, or transformed into something people are willing to exchange in a market. Most users quickly appreciate that there is no free ride in digital networks: we pay for "free" services every time there is an ad on a page. Or as the adage of social media economics goes, *If you are not paying for it, you are not the customer; you are the product being sold.* However, most of us are happy to be such products, given what we perceive we get in return. Participation in digital networks is not coercive in a straightforward manner.

Network Hegemony

If wealth in the digital network is not evenly distributed and participation is disadvantageous, why do we keep participating? In most cases, we might not even be presented with a choice. The college at the State University of New York (SUNY) where I work, for example, made the decision (like many other schools) to accept Google's offer to handle all the school's e-mail "for free." In the face of $410 million in state budget cuts to SUNY in the past two years, it is understandable why public schools are keen to save money wherever they can. And on the surface, getting better functioning e-mail, a full menu of apps (including calendaring), file storage, chat, as well as 2.5 gigs of storage sounds like a good deal. But when I asked whether there would be other options for handling our school e-mail, I was told this would be the only one. As I wrote in our school newspaper,[10] there are reasons why universities like Yale, UC Davis, and Lakehead originally turned down similar deals with Google or, in some cases, filed grievances citing concerns about privacy and academic freedom (although in the two years since I wrote that, all three institutions have switched to Google). For one thing, in these days of cloud computing (where data are stored in remote company servers, not in the user's computer), who gets access to the data is a complex international legal question. If Google stores copies of our e-mail in 3 of its 450,000 servers located all over the world (for data redundancy purposes, which keeps our data safe in the event of a server failure), some individuals at the aforementioned universities had obviously been wondering whether Google is obligated to hand over their e-mails if the corresponding authorities in those countries come asking for them. In other words, if my Google e-mail data and research are stored in Israel or Malaysia, does that give those governments the right to monitor them? But beyond the issue of surveillance by foreign or domestic authorities (in collaboration with Google), my concern is that the decision to switch to Gmail signifies a further privatization of education by effectively putting everyone at our public institution to work for Google, whether they choose to or not. Let us not forget that Google derives 97 percent of its revenue from advertising. And while switching to Gmail does not mean that my colleagues and students started seeing ads for Viagra or teeth-whitening products next to their in-box (Google Apps for Education is ad-free), it does mean that Google is scanning our e-mails and documents to collect more information about us, their users.[11] The more Google knows about us, the better it can sell that information to people who want to target ads at us. The hegemony of networks is insidiously evident in examples such as this one in which participation is presented as a fait accompli, in the absence of options and alternatives, and as an almost naturalized form of commodification in which a social act (sending e-mail to students and colleagues) is almost invisibly transformed into a revenue-creating opportunity for a corporation.

Of course, it is presumptuous to assume that, given a choice, people would opt *not* to use Gmail (most people at my school seemed to think it was a fine idea, or they simply did not care). The fact of the matter is that inequality in the digital network is not experienced as coercive or unpleasant. To the contrary, because it appeals to our egos by allowing us to express ourselves, participation in digital networks is creative and pleasurable. Everyone feels welcomed because there is a place in the network for everyone and everything. Inclusion is the default setting. The inequalities that the network creates are overlooked by most users because the network is perceived as a better provider of opportunities and equality than the alternatives (social institutions or the state, for instance).

The network thus represents a form of hegemony, a system of rule in which a minority can rule over a majority not by brute force or deception but through consensus. From a Gramscian perspective,[12] hegemonic power is predicated on a harmonious relationship between unequal social classes achieved through the formation of a popular discourse of inclusion: political accommodation of the underprivileged allows the ruling class to maintain its privileges by seeming to represent the interests of the ruled. In the context of digital networks, the trope of "total inclusion" establishes hegemony by promoting the idea that the consensual acceptance of the terms of use (which spell out precisely the way in which we are to be ruled) is rewarded by the opportunity to have a presence in the network on the same terms enjoyed by everyone else. The illusory sense of empowerment is further reinforced by the idea that there is no ruling body in the network. This is true to the extent that there is often no centralized authority in most networks. But we could say that the ruler in networks is network logic itself, which specifies the parameters for interaction.

Consequently, participation in digital networks is seen as a productive, beneficial, and enjoyable contribution to the social order (a form of play mixed with labor). In some ways, this paradoxical relationship of the participant to the digital network is reminiscent of the relationship of the colonized subject to the colonial power. As Partha Chatterjee suggests, the colonial project granted the colonized individuals *subjecthood*, although it did not grant them *citizenship*[13] (it offered them a worldview in which they could locate themselves, but it restricted their participation by reducing them to a subjugated role). Likewise, I will be arguing that the digital network can grant participants subjecthood and agency, but because it produces inequality, it also constrains their rights. The network, in short, can only function if members passively adhere to its logic, not if they are actively engaged in questioning it. Hence there is a need to begin to unmap the network, to transcend its determinism through whatever strategies we might devise: obstruction of its growth, disassembling of its parts, localization of its processes, intensification of its virtualities; hence there is a need, in other words, to resist a logic that can only *think* in terms of nodes.

Nodocentrism

While the technological phenomenon is a powerful social determinant, it is also true that humans are responsible for creating and determining technology in the first place. Thus it is probably more exact to say that humans and technologies codetermine each other. However, for the moment let us continue to focus on the fact that network technologies play an important role in shaping our societies, and let us suggest, therefore, that whereas before the network was merely a metaphor to describe society, now it has become a technological model or template for organizing it. A lot of *socializing* happens within the structures and architectures of digital networks (as evidenced by the amount of time we spend interacting with a human being through an electronic screen), but this socializing is shaped by the network in very particular ways, resulting in new ways of experiencing the world.

What I want to suggest is that what we are seeing is not only the pervasive application of the network as a *model* or template for organizing society but also the emergence of the network as an *episteme*, a system for organizing knowledge about the world. To better understand this development, it should be pointed out that the network model and the network episteme serve two different functions: whereas the model is used to design and build actual networks, the episteme allows us to understand the "networked" world, to see everything in terms of networks, and to apply network logic even to things that are not networks. In other words, as social networks are facilitated or enabled by digital technologies, the network ceases to function merely as an allegory used to describe or study particular forms of collectivity. It becomes, first, a technological template for organizing the social; and second, it becomes an episteme or a way to understand and access reality. This episteme not only is facilitated by the technology but also transcends it, becoming a knowledge structure, a way of seeing the world as composed of nodes and links. The shift from metaphor to model to episteme (which will be explored in more detail in subsequent chapters) signals a transition from using the network to describe society to using the network to manage or arrange society, defining the parameters for interaction within the network by prescribing, or obstructing, certain kinds of social relations between nodes.

The most consequential effect of superimposing this technological template and episteme onto social structures is the rendering illegible of everything that is not a node. I call this effect *nodocentrism*. In describing the relationship that nodes have to things internal and external to the network, Manuel Castells writes,

> The topology defined by networks determines that the distance (or intensity and frequency of interaction) between two points (or social positions) is shorter (or more frequent, or more intense) if both points are nodes in a network than if they do not belong to the same network. On the other hand, within a given network, flows have no distance, or the same distance, between nodes. Thus, distance (physical, social, economic, political, cultural) for a given point or position varies between zero (for any node in the same network) and infinite (for any point external to the network).[14]

Thus whereas the distance between two nodes that are part of the same network is finite, the distance between something inside the network and something outside the network is *infinite* (even if, in spatial terms, that distance is quite short). Nodocentrism means that while networks are extremely efficient at establishing links between nodes, they embody a bias against knowledge of—and engagement with—anything that is not a node in the same network. Only nodes can be mapped, explained, or accounted for. The point is not that nodocentrism in digital networks impoverishes social life or devalues what is around us: nodes behave neither antisocially (they thrive in linking to other nodes) nor antilocally (they can link to other nodes in their immediate surrounding just as easily as they can link to remote nodes). The point, rather, is that nodocentrism constructs a social reality in which nodes can only see other nodes. It is an epistemology based on the exclusive reality of the node. It privileges nodes while discriminating against what is not a node—the invisible, the Other.

Nodocentrism does not provide an incorrect picture of the world, just an incomplete one. It rationalizes a model of progress and development in which those elements that are outside the network can only acquire currency by becoming part of the network. "Bridging the digital divide" is normalized as an end across societies that wish to partake of the benefits of modernity. The assumption behind the discourse of the digital divide is that one side, technologically advanced and accomplished, must help the other side, technologically underdeveloped or retarded, to catch up.[15]

The nature and ramifications of nodocentrism can be illustrated with some quick examples.

Search engine results are examples of nodocentrism in the sense that they point to documents, sites, or objects that have been indexed by the network. What has not been indexed is not listed as a result, and it might as well not even exist in the universe of knowable things as far as the search engine is concerned.

Buddy lists, such as the ones used in instant messaging (IM) programs, are examples of nodocentrism because they portray a social network composed of the acquaintances available to chat on that program (even if the friends are currently offline), but they render invisible the acquaintances who are not on the list because they do not use the same program or because they do not use IM.

Nodocentrism is at work in accidents caused by following inaccurate Global Positioning System (GPS) instructions, as when the GPS device tells its user to drive into incoming traffic or a body of water. By relying on the simulated reality of the digital network over the reality of the terrain, humans give precedence to the actuality of the node.

Similarly, when people are pulled from flights because the combination of their names, ethnicities, or religious backgrounds triggers something in a no-fly database, the process of selection of potential threats exhibits a nodocentric logic. The definition of a threat according to its characteristics as a node or its place in the network represents a new way of applying network logic to security.

Algorithmically generated recommendation lists are another example of nodocentrism. These lists might aggregate the opinions of large communities of users, but in doing so, they also operationalize decisions about what is included in and excluded from the list.

We also see nodocentrism at work in the digitization of archives, making analog materials (texts, photographs, recordings, etc.) available online. However, not all materials are digitized, or not all materials are equally accessible to everybody. Nodocentrism can help us talk about the politics of knowledge construction in an age when we seem to increasingly depend on the digital network as a historical archive.

The articulation of nodocentrism and the kinds of inequalities it produces might suggest that the normative goal of unmapping the digital network is to give shape to a noncapitalist information society. However, information, sociality, and capital are entangled today in such a way that to suggest an easy separation would be simply naïve. Furthermore, the spaces of resistance that digital networks have currently opened up, no matter how circumscribed by corporate interests, are important and should not be dismantled just yet. Therefore, it seems prudent at this point to clarify some things about a book that—going by its title alone—appears to issue a call to arms against digital networks. This book will not be arguing that the existence of the digital network, in and of itself, has negative consequences for humanity (I believe that as the designers and users of digital networks, we—not they—are ultimately responsible for what kind of impact they have on our society). Furthermore, the book will not be calling on anyone to stop using digital networks or providing step-by-step instructions for dismantling any kind of digital network. The point is not to embark on a journey to some remote corner of our contemporary life to find subjectivities or sites untouched by digital networks. Thus this book will not be promoting a network Luddism, because no responsible person can afford to be a Luddite. In a world where 1.6 million cell phones are activated every day, inclusion and exclusion from the network are everywhere—embodied not only by the digital divide that separates the haves from the have-nots but also by the digital divides that privilege some sociocognitive spaces and undermine others, or the interior digital divides that separate our networked from our nonnetworked selves. Instead of romanticizing some prenetworked state of being, this book will try to get us to confront the tensions in those digital divides, because the spaces on the "wrong" side of the divide—those not based on the predictable and controllable models prescribed by network logic—will increasingly be considered threats to the network.

So while we need to be critical of the use of digital networks as platforms for participation, I am not calling for a total rejection of the network as a model for organizing sociality or the dismantling of for-profit networks wherever they may be found. Rather, I believe that a reimagining of identity beyond the templates of the network episteme is necessary to articulate new models of participation, and that is what I mean by doing the work of "disrupting" the digital world: unsettling, undermining, and even unmapping what is oppressive in certain structures of thought. This book strives to present a starting point for this kind of unthinking. While some general strategies will be discussed, they will not be presented as subversive tasks intended exclusively for hackers, anarchists, or dissenters. To the extent that we each participate in digital networks, we are *all* already engaged in the production of inequality, and we are *all* also involved in the politics of inclusion and exclusion of the network. Furthermore, no one enjoys absolute inclusion, so we are always already occupying varying states of exclusion. Embodying the organizing logic of the network is part of what we already do, perhaps without even realizing it, and it is the divide between the networked and nonnetworked parts of our identity (the included and excluded parts) that we have to become sensitive to.

While using networks to disrupt networks might make strategic sense at times (what Hardt and Negri call *fighting networks with networks*[16]), the goal of this work is to theorize models that ultimately move beyond network logic altogether. Disrupting the digital network cannot rely only on marginal strategies such as hacking, open-source/open-content paradigms, peer-to-peer sharing, and so on because these strategies rely on the same logic the network does, as I shall argue in later chapters. The challenge is to acknowledge the fact that, since the network is agnostic about what it assimilates and can thus easily extend its reach, there is "no longer a place that can be recognized as outside."[17] This makes the task of being against the network increasingly difficult, since in order to *be against* one needs to occupy a position or framework outside the established paradigm. To Hardt and Negri, this simply means that we must be against *everywhere*—inside and at the same time outside the network (and since every node has limits or borders, the outside is not just what is external to the network but what lies *internally* between nodes). But if nothing is really outside the logic of the network, how can we begin to articulate the ethical and political meaning of being against the network? The greatest obstacle today to the emergence of a critical theory of the network episteme is, therefore, our inability to imagine an outside.

Beyond Networks

In the long term, perhaps more egalitarian organizations might emerge from the process of disrupting or unmapping the network. But today, at this very moment, it is unlikely we can either challenge or substitute the network model if this means reorganizing technological infrastructures and the economy at large. All we can hope for, perhaps, is to reorganize our intimate ways of thinking. If unmapping is unthinking, it should require no special tools or skills but the mind. The present goal of unmapping the network, therefore, is to give the mind the tools to envision how the network has shaped and molded us, to explain how the network has determined us, and more important, to raise the possibility of alternatives—to ask how *we can determine it.*

Perhaps this intellectual exercise is a good enough start, considering that network logic points to a crisis of imagination, specifically, to a crisis of how we imagine ourselves as individuals in a community. Defining the self in relation to the collective requires an investment of multiple desires or affects that converge in the act of imagining a community. In other words, community can be said to be the intersection (whether benign or violent) of affects that start as imagined and, through the process of communication, crystallize into social practices. As Etienne Balibar suggests (in his analysis of Spinoza), it is in the collective process of imagining community that we communicate our desires and work out "the relationship through which affects communicate between themselves, and therefore the relationship through which individuals communicate through their affects."[18] In one way, networks open up new ways for individuals to communicate affectively, giving way to new forms of community and participation. But as has already been suggested, the network determines those forms of community according to specific interests. We might be fascinated by the digital network as a new form of imagined community,[19] but we need to ask, *Whose* imagined community?[20] Who is doing the imagining, and who is merely living in the product of someone else's imagination? If hegemonic power is inscribed in networked communities, we need to ask *what* the network template leaves for us to imagine, which is why the network template represents, to paraphrase Chatterjee,[21] a colonization of our collective power to imagine community.

It is, in fact, the very appeal of the digital network as a cultural metaphor for imagining community that makes it particularly restrictive as a social determinant. The digital network is a ready-made image into which we can pour our hopes for social unity and connectivity. We can point to a location in the network map and say "that's me!," while admiring the wealth of our social capital. A network map thus becomes an egotistic object for aesthetic contemplation: it is visually pleasing, dynamic, and it is *about us.* It is the social world turned into an interactive mirror, miniaturized and projected onto a screen for our pleasure. The digital network signifies the aestheticization of the social, a means for the masses to contemplate a simulation of themselves and express themselves through this simulation. But it also represents an arena of restricted or diminished opportunities for meaningful political and social action. Walter Benjamin had already described similar dynamics in relation to Fascism. According to him, the emerging Fascist rulers recognized and feared the potential of the masses to change property relations; in order to preserve the traditional property system, Fascism found its salvation "in giving these masses not their right, but instead a chance to express themselves"[22], thus introducing aesthetics into political life. In other words, Fascism granted the masses subjecthood, the ability to express themselves, as a way to avoid granting them more active powers. Interestingly, for Benjamin the aestheticization of the political involved the masses accepting *reproduction* (what we would call *simulation*) in lieu of the "uniqueness of every reality."[23] Likewise, today's networked masses are encouraged to express themselves in a simulated social sphere that contributes to the reproduction of inequality. They are encouraged to accept the network map in place of the "banality" of unnetworked space and to express themselves through it. The rendering of politics as aesthetics satisfies the need for sociality while respecting the traditional forms of property on which capitalism is founded.

Thus far, there has not been a widespread movement to challenge the hegemony of the network and its colonizing imaginary. Hegemonic rule depends on widespread consensus, which in network terms means all nodes subscribe to the same protocols and accept the same models of social participation. Public intellectuals (media gurus, academics, etc.) who advocate that digital networks are being used to empower the public are only undermining our potential to free ourselves from the hypnotic hold of this aestheticized form of sociality. This is why there is a need to theorize how new imagined communities can be different from the template-based communities of the digital network. At the same time, any alternative would have to organize itself in order to survive, and that form of organization would probably *look and act* just like a network. While I am attempting to critique the network as a digital template for sociality, I also recognize that the network, as an organizational form, can be useful. If the only way the excluded can unsettle network hegemony is to first organize themselves into a *networked* multitude that eventually rejects, subverts, or disinvests itself from network templates, so be it.

Unmapping the digital network needs to involve *both* working within the spaces of resistance that digital networks have already made available *and* asking what it means to obliterate those very spaces.

This brings me back to the project of imagining and thinking alternatives, right here and now, using the digital network as a starting point. Digital networks map unto a social domain what was before unimaginable, reorganizing the possible. They are the result of previous social models as well as new, emerging ones. This actualization of the virtual unveils new associations, new ways in which things that were not linked before are now related, and also in which other things are now excluded or forgotten. Disrupting the network prevents the energy of nodes from becoming arrested or complacent, and unleashes it in new directions, as nodes begin to unthink themselves. From the perspective of the node, the witnessing of the ethical resistance of the outside (the way it is excluded, the way it resists assimilation) can lead to the kind of self-questioning that can generate personal and social change. Sensing the limits of nodes within and outside us can lead to the alteration of our intimate ways of knowing the world through an increasingly dominant corporate nodocentrism. It is ultimately about changing the way we understand others and ourselves.

Thus while this is a book about thinking and unthinking networks, it is also a book about alterity and othering—about the way we imagine and engage *difference*. Specifically, it is about the ethics of othering. In the standard view of interaction in a network, we have two or more nodes struggling to communicate in the presence of noise, as depicted in the Shannon–Hartley theorem, which calculates the maximum amount of data that can be transmitted given a specified bandwidth and noise interference. Noise—we have always assumed (at least since Shannon's "Mathematical Theory of Communication" was published in 1949)—is a barrier to interaction, and this model has influenced our development of communication theories and technologies. The Internet, however, practically eliminated the problem of noise through digitization and packet switching (distributing information in small chunks through multiple channels). But the project of unmapping the network asks if we have perhaps invisibilized noise too quickly and too efficiently. Noise, in network terms, is nonnodal—it is not simply a meaningless sound but a sound that does not conform to the harmonies of the network. The project of disrupting or unmapping the network and encountering its outsides is one that goes from trying to solve the problem of *communicating in the presence of noise* to one that sees *noise as communicating presence*, the presence of the Other. In short, noise communicating difference. It is only in the outside spaces of the network, beyond the limits of nodes, where we can acquire enough clarity to listen to the sounds that alternative subjectivities, even from within us, might suggest.

2 THE PRIVATIZATION OF SOCIAL LIFE

IN HIS BOOK *The Wealth of Networks: How Social Production Transforms Markets and Freedom,* Yochai Benkler suggests that the information economy has ushered in an era of human cooperation in which the limits of capitalism are transcended by new models of social production, facilitated to a large extent by digital networks.[1] These open, commons-based peer production models (which challenge the old economic models) position humans not in the traditional role of competitors in the market, but as collaborators in a social environment. According to Benkler, in these networks "a good deal more that human beings value can now be done by individuals who interact with each other socially, as human beings and social beings, rather than as market actors through the price system."[2] Unfortunately, many of the authors who write about the digital network tend to bypass the issue of who owns and controls it and for what purpose. This is an important matter to consider if we want to formulate a comprehensive critique of the digital network, for it can help us move away from simplistic questions about whether we should use the network or not to more relevant (and more difficult) questions about the kinds of relationships we enter into when we use digital networks. Much like my university's Information Technology department, we will increasingly find that we cannot avoid using the free and efficient products and services provided by companies like Google, Microsoft, Facebook, and so on. What does this mean for us, the public, and for alternatives inside and outside the network?

On a short blog post made on March 10, 2010, and appropriately titled "Bike Maps: Triumph of Corporate Solutions over Grassroots?," Charlie DeTar[3] reflects on the significance of Google's launch of a function for Google Maps that lets the user calculate bicycle routes. Up until that point in 2010, interactive bike maps were available online thanks to various grassroots communities of environmentally minded programmers and enthusiasts with a do-it-yourself attitude who gathered together and—using open source software and crowd-sourced data—put together services like Bikely.com and Opencyclemap.org. Many of these websites were real examples of peer-to-peer distributed models of collaboration. They were not perfect and their coverage was relatively poor (consisting only of the areas that members of the community were interested in mapping), but they represented the spirit of collaboration and entrepreneurship that characterized the open-source movement. Then in one swift move, Google decided to apply its engineering expertise, capital resources, and mapping infrastructure to provide bike maps for 150 cities around the world, making the grassroots solutions practically obsolete in the opinion of many. Of the lovingly constructed grassroots sites, *Wired* magazine dismissively remarked, "No longer do [bikers] have to rely upon paper maps or open-source DIY map hacking."[4] To be sure, Google's service will benefit users who are not currently reached by the grassroots bike route sites (these users will now get a "free" service without having to hassle with learning the skills necessary to participate in an open-content project). And the grassroots sites, with their devoted communities, will hopefully not disappear overnight. But will new grassroots sites emerge to compete with Google now that it has entered the market? How long will the existing grassroots sites continue to thrive? What will be their motivation to innovate? Does the dominance of the corporate solution matter?

These questions are obviously not relevant only to bikers and their maps. More and more, we see individuals—even those whose vocation is to remain critical of capitalism—grant corporations more control over their content and their privacy. The result is a system that compels the public to participate mainly because of the perceived benefits of having their data hosted and distributed by the network with the most number of users. How this benefits the profit margins of corporations is obvious, of course. But what does the public get in return?

Communicative Capitalism, Commodification, and Inequality

A useful point of departure for explaining how digital networks generate inequality through participation is the concept of communicative capitalism. Jodi Dean defines communicative capitalism as "the materialization of ideas of inclusion and participation in information, entertainment, and communication technologies in ways that capture resistance and intensify global capitalism."[5] In communicative capitalism, everyone has the tools and opportunities to express an opinion. "Participation" in society is therefore identified first and foremost as the ability to communicate, to express one's opinion, in particular about the—mostly commercial—choices that give individuals their identity. For instance, if I prefer Google's Android platform over Apple's iOS, or Republicans over Democrats, I see it as my duty to express this opinion and to express it frequently. Consequently, the overabundance of communication in a marketplace in which all opinions compete for visibility results in an *everything goes* kind of democracy where change is impossible (after all, if all options are equally valid, how can one course of action be declared superior?). Challenges to the status quo are thus ineffective, as any resistance to capitalism is diluted as merely another option, another alternative in the marketplace of ideas. The only thing that endures is capitalism itself.

In this context, networked participation itself can be narrated as an expression of the spirit of capitalism[6]: it is fair (contributes to the common good), it promotes security (contributes to the well-being of the economy and therefore our well-being), and it is exciting (it offers liberation through new opportunities for growth). The more we participate in digital communication networks, the more this ideology is reinforced. To paraphrase Deleuze, communicative capitalism does not stop people from expressing themselves but forces them to express themselves continuously.[7]

Communicative capitalism means that communication and social exchange take place not just in any environment, but in a privatized one. In essence, the neoliberal impulse to subsume all social communication and participation to market forces can only be achieved if the network is made the dominant episteme or model for organizing social realities. This is accomplished by the application of a nodocentric filter to social formations, which renders all human interaction in terms of network dynamics (not just any network but a digital network with a profit-driven infrastructure). Under this nodocentric view, the goal is to assign to everything its place in the network. Thus to be anything other than a node is to be invisible, nonexistent. The technologies of communicative capitalism are applied toward the creation of a pervasive or ubiquitous computing environment in which every thing and every utterance must be integrated or assimilated as a node in the digital network.

The argument that digital networks have paved the way for this new "participatory culture"[8] requires us to accept the premise that the continued privatization of the public sphere is the best avenue for social exchange, cultural production, and civic engagement. Notwithstanding experiments in open-source software, peer-to-peer (P2P) file sharing, and so on, digital networks are, at some point or another, for-profit ventures (at best, "open" movements can only sublimate or delay commercialization). And while the performance of public acts in private venues need not imply exploitation or oppression (a privately owned newspaper can still provide an important public function, as can a café in which people gather to converse), the difference is that while digital networks do increase the opportunities to act and participate, they also exploit the gap between network participants and those who profit from their aggregated contributions. For reasons that will become clear in this chapter, the exchange between participants and network owners is not symmetrical or fair. For a digital network to operate successfully and support "free" participation, it must figure out a way to exploit the creative and social labor of participants and turn that participation into a commodity, into something that can be exchanged for capital. Thus while it is true that the technologies of communicative capitalism embody practices of inclusion, they also perpetuate the ideology of capitalism and obstruct any resistance to it, as Dean proposes.[9] Particularly, they increase inequality through commodification, the transformation of social activity into a commodity that can be bought or sold.

Commodification is a concept from Marxist theory that refers to the process of taking something that is outside the market (something without commercial value) and bringing it into the market, turning it into a commercial transaction. What was previously exchanged or supplied freely is now part of an economic exchange, which reduces its worth to a material value and opens up opportunities for exploitation. If, for instance, people used to share recipes with each other at social gatherings, but now they do so through a website operated by a corporation, one could say this action has been commodified. Or if an individual is engaged in a new kind of cultural activity that can only take place on a for-profit digital network (e.g., sharing digital videos), then this is also an example of a commodified social act.

There are three simple examples of how commodification works as a process within capitalism. The first one is privatization, where services (such as education, health, transportation, etc.) provided by the state are replaced by services citizens have to pay for out of their own pocket while continuing to pay taxes. The second one is commercialization, where things like scientific research increasingly serve private, not public, interests, or where intellectual property laws keep cultural goods in private hands for longer instead of releasing them as public goods. The third example—which is the one most relevant to our discussion of digital networks—involves the socialization of labor. The easiest way to understand socialization as an instance of commodification is to think of women's labor in industrialized nations. In this context, socialization of labor has meant taking certain domestic tasks traditionally performed by women in a patriarchal society (such as cooking, cleaning, child rearing, etc.) and converting them into activities that one can pay someone else to do, or developing products that make those tasks easier. This process of commodification has allowed women in industrialized countries to escape domestic servitude and enter the workforce. The opportunity to be exploited as workers might not seem like much of an improvement, but to some it represented a step forward because it afforded women certain benefits, like the opportunity to become more independent by earning their own money, the opportunity to organize themselves in unions that challenge exploitation, and so on. One could critique this rather simplistic account of capitalist processes by pointing out that if, indeed, the socialization of labor has empowered any women workers in industrialized nations, capitalism has responded by exploiting women of color elsewhere. But the claim is that without this process of commodification, which grants more freedom and independence to some types of exploited workers, there would be no eventual challenge to private property and no eventual breakdown of capitalism.

The question is whether the commodification of the social that is inherent in digital networks can indeed eventually lead to a means of resisting the inequalities that capitalism produces, or whether it merely contributes to their entrenchment. The answer to that question is, of course, something that needs to be continuously readdressed at every site and at every moment in history. But while it is not easy to establish exact parallels between the commodification of women's labor and the commodification of sociality in digital networks, some analogies can be drawn in regard to how both forms of commodification can be experienced as alienating and dehumanizing in certain respects, while at the same time empowering and liberating in others. In other words, the commodification of the social in digital networks, the process whereby our social lives are subordinated to the logic of nodocentrism, can both open and close productive forms of sociality that challenge capitalism.

One way to talk about these contradictory effects is to talk about the dual processuality, or double affordances, of networks. As Jan van Dijk[10] observes, networks make two sets of outcomes possible at one and the same time: a scale expansion accompanied by a scale reduction, more freedom of a certain kind but more control of another, more openness at one level but more constraints at another, and so on. Alexander Galloway describes a similar tension between two opposite but complementary dynamics that play out in the protocol or code of digital networks: one that "radically distributes" control and another that "focuses control into rigidly defined hierarchies."[11] The double affordances in digital networks make possible dual processes to be present at once, which is why the commodification of the social might look very differently depending on which angle one is looking at it from.

For instance, participants in the digital network may experience a high degree of freedom when it comes to deciding what groups to form, what content to create, and so on; on the other hand, corporate power seems to curtail that freedom, as corporations retain control over which new features to implement in the network, which members to expel, or even whether the network will continue to exist in the future or not. Likewise, increased opportunities for content production are countered by the transfer of property rights to the corporation, as happens when corporations acquire the intellectual rights of whatever content users create and upload to the network. In another example, the diversity of voices found in multifaceted communities of interest is countered by the homogenization of software platforms, which means that all communities must use one set of tools and abide by one set of rules: the corporation's. This dual processuality helps explain why it is difficult to make quick pronouncements about the positive or negative effects of the commodification of the social in digital networks. Alternative practices are always possible, even if they are quickly assimilated into the larger organizing logic. But dual processuality does not help us explain why, even when the effects of commodification are perceived to be largely negative, people keep participating in the network.

Participatory Culture and the Society of Control

Participation in digital networks is rewarding. It is both a form of labor and a form of play—or *playbor*.[12] It is an activity that appeals to our superego, an imposition by an authority that "enjoins one to enjoy"[13] rather than forbidding enjoyment. But while it is play, it is not an unconstrained, free-form type of play, the kind that is chaotic and unplanned, full of possibilities. Rather, it is a rationalized game, standardized and institutionalized, that contributes in very specific ways to a capitalist social order.[14]

This rationalized game is very much dependent on the mechanics of exclusion and inclusion of the network. In order to play, what is outside the network must be assimilated and brought into the network. This form of playbor is freely and enthusiastically performed by those already inside (which is why invitations to join the latest social media craze are more effective when they come from a friend, not a company). Once inside, players encounter a hierarchy between those new nodes with few links and those super-rich nodes or hubs, which everyone keeps linking to. The game then becomes trying to acquire as many links as possible, in an attempt to approximate the status of a super-rich node.

Participation is thus both a form of violence and a form of pleasure. More than a desire, participation is an urge, a form of coercion imposed by the system. This logic is internalized, rationalized, and naturalized. Participation in the network is a template for being social, for belonging. It is perceived as socially rewarding. It gives the illusion of making us more social. In the disciplinary societies of the nineteenth century, the self was actively molded into conformity by institutions external to the body, like the factory or the school.[15] The participatory culture of the digital network has more in common with the society of control, where the desire to conform emerges from within the body.[16] By setting the parameters for inclusion, the network episteme perfectly expresses this new architecture of power. No external institutions are required to enforce this episteme because it is affirmed through our personal use of technology, establishing the network as the main template for organizing and understanding the real.

Because digital networks have many participants, it would appear as if ownership of the network is distributed. But in reality, what is distributed are the opportunities for generating value for the companies that own the various parts of the network. This work can be done by anyone, anywhere. Labor is no longer conducted at the workplace in exchange for a wage. Rather, it is produced mostly outside the workplace, during our "free" time. It is rewarded not with a paycheck but with social capital such as attention, rank, and visibility. Surrendering privacy and property, lured by promises of fleeting viral fame and motivated by fear that we will be the only ones left out, the urge to participate impels us to upload the fruits of our creative labor and hand over the social capital of our electronic address book.

This is a form of participation that transcends labor. It is the privatization of social production, of the creative cooperation that happens when people interact to give shape to new cultural forms. Companies have recognized this as a business opportunity: the appropriation of the free labor of socializing and its reinsertion into the market as a commodity. Under the pretense of creating communal gift economies in cyberspace, social beings are put to work for corporations. And while there are attempts to protect creative social labor under new collective forms of ownership or "peer property" (licenses such as GNU, Creative Commons, etc.), the fact is that these models cannot escape commodification at some level or another (one might be able to release content under an "open" license, but it still needs to be distributed over the wires and technology of a "closed" infrastructure, as further discussed in chapter 6).

Some might ask, is the expropriation of our playbor a small price to pay for the emerging forms of sociality that digital networks make possible? New modes of production and avenues for organizing action do in fact emerge, but they become arranged under a structure where every aspect of the public is owned, hosted, or powered by private interests. A quick look at the terms of use of any Web 2.0 company will reveal as much. Thus playbor continues a trend where—to paraphrase Frédéric Vandenberghe[17]—the social is increasingly subordinated to the economy. As reasons to opt out become harder to rationalize (nobody wants to be an outcast; these days, even antiestablishment dissenters have Facebook profiles), the public sphere devolves into a privatized peepshow, where every contribution to the commons cannot escape commodification, and where user-generated content is valued not in terms of its quality, but in terms of its potential to be mined for information that contributes to the maximization of profit.

Some authors have begun to wonder about the limits of a participatory culture in the context of capitalism and consumerism. Peter Levine,[18] for example, discusses the challenges that students face and will continue to face in finding appropriate audiences for their civic-oriented participatory media work in an environment dominated by commercial products. As Kathryn Montgomery[19] also points out, despite the numerous examples of youth empowerment with digital media, important questions remain about whether these new models of participation can be adopted by larger segments of the population and applied to a range of issues outside of high-profile events such as national elections. She observes that "the capacity for collective action, community building, and mobilization are unprecedented. But the move toward increasingly personalized media and one-to-one marketing may encourage self-obsession, instant gratification, and impulsive behaviors."[20] Likewise, Stephen Coleman[21] questions the capacity of government-driven digital media curriculums to address questions that might potentially challenge the power and legitimacy of corporations or the state. His work serves to remind us that the models of participation that technology affords are shaped to a large extent by the politics of the institutions that make the technology available.

A common thread in most critiques is that authority in the participatory culture operates not by threatening to expel us from the network, but by making it difficult to resist participating in the network in the first place. The more one participates in digital networks, the more totalizing this form of authority becomes. We are impelled to use certain services ("you *must* join this site; all your friends are doing it!"), submit to their terms of use, and accept the barrage of advertisement while pretending we can ignore it. This is a form of "friendly violence that doesn't appear violent at all."[22] In fact, it looks and feels positively prosocial. Perhaps that is why there is such an emphasis on amicability in social media (friending, liking, etc.), to conceal the "friendly violence" of a form of participation that undermines the public interest and obliterates alternatives.

The network episteme reinforces a narrative where participation is productive, while nonparticipation is destructive. Within the network, everything. Outside the network, nothing. *All* forms of participation are allowed, as long as they submit to the organizing logic of the network. Participation itself then becomes the only means of expressing difference. By adopting this logic, however, we reject the forms of difference and disidentification that are achieved through nonparticipation. Thus the belief that participation in networks creates equality and diversity is, in fact, a rejection of difference, because ways of belonging that do not conform to nodocentrism become an impossibility within the network.

Capitalizing the Social

In a popular article, "Is Google Making Us Stupid?," Nicholas Carr[23] argued that the Internet is diminishing our powers of concentration, taxing our attention with advertisements, and promoting a broad but superficial kind of knowledge that erases the possibility of a shared cultural meaning. Of course, he targets Google because of the company's dominant, although by no means exclusive, role in turning information into a commodity and wanting to supplement—perhaps eventually even replace—our brains with a kind of artificial intelligence that can process information more efficiently. Although far from being a radical anticapitalist, Carr's point in critiquing Google is that much is at stake over who gets to define what the models of information processing look like. This is a point that can also be made about some corporations' influence in defining emergent models of social organization. If Google is changing our cognitive makeup, Facebook is rewriting our social one. The rise of the digital network as a template for organizing sociality means that corporations are playing and will continue to play a major role in shaping the modes of participation and citizenship in our societies. To better understand the implications of this process, we can look at the technologizing of society through the economics and market structure of the social networking industry.

Social network services such as Facebook and MySpace are web-based platforms that allow users to create a personal profile by filling out a form that collects personal information. Once a profile has been created, the user can "friend" other users by linking to their profiles. Users can also become members of various groups that share similar interests.[24] Social network services can map already existing networks (for instance, a group of students taking a college class) or they can map new networks of people who were previously unconnected but who are brought together by a common cause (e.g., a local, national, or global group supporting a social cause).

Encouraging the compulsive and continuous expression that communicative capitalism thrives on has turned out to be a profitable business model, as evidenced by the growth of the social media industry. Facebook, launched only in 2004, was adding on average 250,000 new members a day by 2007. Currently, it has more than one billion members, who perform more than 60 million status updates everyday and share 30 billion pieces of content every month, as cited in data posted on the statistics page of their website. According to industry reports, the online social networking market as a whole grew 87 percent from February 2006 to February 2007, accounting for 6.5 percent of all Internet visits.[25] During roughly that same window of time, MySpace grew from 66.4 to 114.1 million users, Facebook went from 14.1 to 52.2 million members, and Orkut (owned by Google) from 13.6 to 24.1 million members.[26] Social media are driven by advertisements targeted to users based on the demographic data they provide, and the amount spent on advertising in social network services was $1.4 billion in 2008, with companies spending $305 and $850 million to advertise their products on Facebook and MySpace, respectively.[27]

It is a booming, if volatile, business. But while the issue of who owns the social media determines, to a large extent, the experience of the user and the opportunities for participation available to her, the question of corporate ownership often gets overlooked because there is a widespread perception that these new technologies are increasing civic participation, regardless of who owns them. For instance, "The Internet and the 2008 Election," a study by the Pew Internet and American Life Project, reported that 46 percent of the population used the Internet, e-mail, or text messaging to "get political news and share their thoughts about the [U.S. presidential] campaign."[28] Although, as expected, the larger portion of that figure is composed of people who simply use new media to receive or retrieve information, the study reports that around 11 percent of the population of the United States actively used those tools to contribute to the political conversation by forwarding or posting someone else's commentary about the race. Specifically, 5 percent of the population posted their original commentary or analysis to an online news group, website, or blog.[29] It should come as no surprise that young people are leading this trend, and one of the tools they are most likely to use for this purpose is a social network service. About two-thirds of Internet users under the age of thirty have a user profile in a social networking website like Facebook or MySpace, and according to the Pew report, about 40 percent of them have used these sites to engage in political activity of some kind.[30]

It is undeniable that social network services provide some opportunities for social and civic participation. Since their use is increasing (as of 2010, the world spends over 110 billion minutes a month on social networks and blog sites, which equates to 22 percent of all time spent online[31]), we would expect to see a more socially and civically engaged population. Even if such a population is emerging, and we dismiss criticisms that digital networks only promote the kind of "slacktivism" that supports feel-good causes with little impact, there are still not a lot of questions being asked about the kinds of privatized environments in which civic and social participation unfolds.

On the one hand, then, we see an increase in the use of social networking services. Most of the research cited previously seems to suggest that a growing portion of the population (especially the youth) will continue to use social network services to engage in some form of social participation. On the other hand, we must also acknowledge the fact that the most popular of these social networking sites are privately owned. There are, indeed, examples of noncommercial social networking services; but when compared to the millions of users of for-profit social network services, it is obvious that they cannot compete with them in terms of popularity and reach. It is the commercial nature of social network services and its impact on new forms of social organization and participation that concerns us here. There is no denying that corporations are responsible for most of the innovation we are seeing in social networking services. The question is about which designs become dominant, and what forms of social participation they normalize.

When looking at traditional forms of media like television or radio, we usually distinguish between corporate and public providers because we believe the issue of ownership makes a difference in terms of mission, objectives, social obligations, use of advertising, view of audiences as consumers or citizens, diversity of voices, transparency, attitudes toward regulation, and so on. But curiously, even those researchers who see social networking technologies as advancing more active forms of citizenship have mostly neglected the question of how these forms will be actualized under the corporate models that most users will be exposed to. Missing, then, is a discussion of how the commodification of the social gives way to a particular market structure where digital networks are controlled by fewer and fewer corporations, and how these corporations acquire and redistribute user-generated content in a way that undermines a democratic constitution of the public sphere.

The Dominant Market Structure of Participatory Media

The mass adoption of corporate-owned digital networks has somehow been heralded as the end of cultural monopolies. Power has shifted, we are told, and no longer is an elite minority in control of the production and dissemination of messages. That capacity has now been distributed among a new army of content producers who digitize, analyze, aggregate, and share content without a need for permissions or licenses, and who face no steep barriers of entry. This new state of affairs is summarized in Jay Rosen's manifesto, "The People Formerly Known as the Audience," in which New Media says to Old Media, "You don't control production on the new platform, which isn't one-way. There's a new balance of power between you and us."[32] No longer are we dependent on a handful of broadcasters, publishers, or studios, apparently. Now *we* are the media, and our ranks are made of citizen journalists, blogger mommies, Wikipedia editors, garage bands, eyewitness videographers, mobile activists, consumer reviewers, self-published pundits, and so on. In this democratic agora, experts, and gatekeepers have been supposedly replaced by a smart mob of amateurs, a crowd supposedly wiser than any single expert. Instead of information flowing one-to-many, now it is generated and distributed in peer-to-peer fashion, many-to-many. This revolution in cultural production has, in theory, ushered in a new era of equality and creativity, a utopia where all participants have the same opportunities and where they voluntarily and freely cooperate with each other in the production of common goods that can be shared by anyone, replacing top-down hierarchies with open modes of production where cooperation and reciprocity are more important than the generation of profit. Subscribers to this idealistic discourse of *digitalism*[33] believe that the Internet can be a space free of exploitation, and that the new models of cooperation are leading to the only realistic alternative for reimagining the failed social institutions of our times (the state, the corporation, the school, the church, etc.).

Unfortunately, the immense promise of these new models of interactivity has somewhat obscured the fact that more and more aspects of this public sphere are controlled by private interests. The Internet has become almost completely subordinated to the forces of the market, and while users gain access to services and tools cheaply or even "for free," they do so at the cost of being exposed to a barrage of advertisements and having their every movement within these networks tracked and logged. From a neoliberal standpoint, this might not seem like a problem. The privatization of social space is in fact something to be encouraged because markets are seen as engines for democracy. Thus corporations—not governments or civil society—are believed to be best equipped to meet the communication infrastructure needs of democracies; they are optimally positioned to supply low-cost and innovative technologies, providing citizens with more opportunities to generate opinions (not just receive them) and increasing their ability to respond immediately and effectively in the public sphere.

But the conflation of markets and democracy is not, as we know, without its (rather serious) problems. For one thing, production in a market tends to be oriented toward what sells, not necessarily what is best for society. Second, markets tend to display undemocratic power differentials because one dollar, not one person, equals one vote.[34] In other words, not all actors in a market have the same power or access to the same resources. The so-called open or flat markets of the information age replicate these failings to a large extent because these markets where supposedly all participants are equal are not free of exploitation; they are built with devices, products, services, and knowledge structures that—to various degrees—replicate exploitative dynamics.

To cite but a few examples, consider the conditions of near or actual slavery under which Coltan (columbite–tantalite), a mineral contained in most of the electronic devices that power the "free" Internet, is mined in the Republic of the Congo. Profits from the mining of this mineral have financed war, rape, and murder in Africa.[35] Or consider the suicides at the Foxconn factory in China (a manufacturer of components for Apple products), which are said to be the result of working conditions and pressure from managers (workers there are "reduced to repeating exactly the same hand movement for months on end"[36]). Or consider also the devastating effect that our twenty to thirty million tons of yearly electronic waste (discarded laptops, phones, printers, and so on) is having on countries like China, where ill-equipped recycling centers contaminate the environment and increase the rates of cancer and cardiovascular diseases.[37] Should our digitally augmented democracy at home be built on the promotion of oppression, exploitation, and pollution somewhere else?

But let us continue to explore this claim that the one-to-many monopolistic model of communication has been replaced with something more democratic. At a superficial level, of course, it has: instead of a handful of voices, there are many. But what has the monopoly been replaced with? In this era in which users—not monopolies—generate content, users must still make decisions about which tools to use to distribute their content. If, for instance, someone has captured the antics of an adorable cat on video, and that person wants the video to be seen by the largest possible audience, she or he will think immediately of one place to upload the video: YouTube. Similar decisions will drive users to satisfy their social networking, microblogging, or photosharing needs by going to Facebook, Twitter, and Flickr, respectively. To be sure, Flickr (owned by Yahoo!) has some competition from Picasa (owned by Google). But the market is still dominated by only a handful of choices.

Thus at a time when user-generated content supposedly rules, the single-seller monopoly has merely been replaced by the single-buyer monopsony. A monopsony, in economic terms, represents a type of market structure where many sellers encounter a single buyer (as opposed to a monopoly, where one seller has many buyers). The monopsony, I argue (or oligopsony, if there is not just one but a few competing buyers), is emerging as the dominant market structure of the digital network. If users want their content to be easily accessible (or have a chance to go viral), there is only one place to go sell or, in most instances, surrender their content: large companies like YouTube, Twitter, and so on. Thus one-to-many is not giving way to many-to-many without first going through many-to-one.

That monopsony has become the dominant market structure of the web is not accidental. The architectures of participation of social media are based on a model where profit margins are maximized the more users join the network (which is why access is free or extremely low cost), and the more demographic data those users provide so that advertising can be targeted at them. In other words, if we are not paying for a product, we *are* the product. Access to free social media services exist only because companies have figured out a way to monetize our participation.

The Economics of Media Conglomeration

In certain segments like social media, the launch of new companies (there seems to be a handful of start-ups emerging every week) gives the impression of a competitive market. But merger and acquisition trends suggest a move toward conglomeration that mirrors that of (and intersects with) traditional broadcast media. In a notable example, MySpace (which currently has over 185 million members) was acquired for $580 million in 2005 by Rupert Murdoch's News Corporation, one of the eight companies that dominate the global media market[38] (the fact that six years later MySpace was sold by News Corporation for only $35 million has more to do with changes in the market and does not signal a diminishing trend in corporate conglomeration).

Historically, media that depend heavily on advertising to generate revenue tend to become larger and larger conglomerates. Bigger audiences mean more eyes to sell to advertisers, so a surge in participation represents an increased opportunity for generating profit. This is the reason media corporations seek to eliminate competition and acquire ever-larger audiences. It is the same logic that dictates why a small city cannot have two major newspapers: too many newspapers in one city would mean that the advertising revenue pie is sliced too many times and profit margins for each media outfit become smaller, making it impossible for the media firm to operate. Alternatively, with only one single newspaper dominating the market, profit margins are bigger and the newspaper is better able to fulfill its social mission by paying reporters and staff competitive salaries. This has been the reasoning behind the special regulatory dispensations made in favor of the media industry.

While websites are not newspapers, it is interesting that the same argument is used to excuse anticompetitive behavior when it comes to monopsonies. The question of the government's role in allowing these "natural monopolies" to thrive in the United States deserves some consideration, especially because it is necessary to correct the misconception that only one political party is interested in helping media corporations become bigger and more profitable monopolies (in turn giving them unprecedented political power). The truth is that for decades—and under both Republican and Democrat leadership—the Federal Communications Commission (FCC) has pursued an agenda of active deregulation that has allowed a handful of media companies to acquire more and more market power.[39] Media formats might change, but the practice of protecting corporate interests has continued, even if it is under a populist doublespeak. Under the current Obama administration, for instance, calls by the FCC in favor of net neutrality (hinting at possible regulations that would ensure transparency and corporate accountability) already advance the notion that Internet users are to be conceptualized not as active citizens who are entitled to commercial-free public space, but as passive consumers who are merely spectators in the theater of deregulation, a process supposedly carried out for their benefit. In other words, net neutrality, as envisioned by corporate and government interests, is a euphemism for some degree of transparency while deregulation and conglomeration continues as planned (more in chapter 6).

A free market in which competition really drives innovation exhibits advantages not found in a system of centralized control and regulation, which is precisely why the Internet works well when it does not get too bogged down by restrictions. But the enabling of powerful monopolies/ monopsonies through deregulation creates less, not more, competition. This ultimately is a disservice to the public. Digital networks have become important public spaces, and it is crucial to ensure that the system remains competitive.

A careful analysis of the ways in which capital and sociality are entangled in digital networks is important for another reason. As trends toward the privatization of social spaces continue, the expression of what is considered outside the norm will become possible only in unnetworked spaces, away from the participation templates of the monopsony. Disidentification—imagining and claiming difference in opposition to the digital network monopsony—will become a necessary step in the actualization of alternative ways of knowing and acting in the world. But before discussing how this might be possible, a better understanding of digital networks as models for organizing the social is necessary.

3 COMPUTERS AS SOCIALIZING TOOLS

IN HIS BOOK *Images of Organization*, Gareth Morgan proposes a way of understanding organizations through the metaphors employed to describe them. One can imagine organizations, he argues, as machines that process inputs and outputs, organisms that interact with their environments, brains that learn from their experiences, cultures that enact the shared reality of their constituents, political systems that manage conflicting interests, or psychic prisons that impose restrictions on our actions and thoughts. Each one of these metaphors provides us with a different vision of what social organizations are, what our role within them is, and how they should be managed. If the book had been published a few years later, after the Internet and other digital technologies had become part of our lives to the extent that they have, Morgan would no doubt have had to consider what is now perhaps the dominant metaphor for describing social systems: the network.[1]

A network is a system of linked elements or nodes. It is a concept that can be used to describe and study all sorts of natural and social phenomena. In fact, the concept of the network has become such an abstract trope that it can be used to describe almost everything that consists of two or more associated entities. For the present purposes, however, we are concerned primarily with digital technosocial networks. Digital networks, to reiterate, are social systems linked by digital technologies. Borrowing from a standard definition of a social network that is mediated by some form of computer technology,[2] we can broadly define a digital network as an assemblage of human and technological actors (the nodes) linked together by social and physical ties (the links) that allow for the transfer of information among some or all of these actors. Digital networks are complex structures reflecting in some measure each of the metaphors described previously by Morgan: they are part machine (their backbone is digital information and communication technologies), part organism (they are powered by the actions of living beings), and part brain (the combination of people and machines produces a form of collective intelligence that is, supposedly, greater than the sum of its individual parts). Obviously, these networks also easily illustrate the metaphors of cultural and political systems. And depending on who you ask, they either exemplify the metaphor of the technocratic psychic prison or represent the only sustainable and scalable alternative to revitalizing what some describe as the failed institutions of our times.

But unlike Morgan's images, networks should not be treated just like any metaphor. As Galloway[3] argues, using the network as a cultural metaphor to signify notions of interconnectedness is limiting and misleading, given that the network in our age is not just a metaphor but a material technology that is a site for concrete practices, actions, and movements. And this is key to understanding the impact of digital networks: as social networks are enabled by digital technologies, they become templates or architectures for organizing the social. If the network was a useful metaphor to describe society before, now it has become a (for-profit) model or architecture for structuring it. Eventually, this architecture becomes an episteme—a way of organizing our theories about how the world works. As pointed out earlier, the shift from metaphor to model to episteme signals a transition from using networks for describing society to using networks for managing society, facilitating or obstructing certain kinds of knowledge systems about the world.

Given that digital technologies are cybernetic technologies, to talk about networks as templates is to talk about networks as computer models. The technological part of digital networks is made up of computer code or algorithms. To be sure, these algorithms do not appear out of thin air. They operationalize behaviors according to their author's understanding of how society works, frequently informed by a relatively new branch of science known as network science. In conjunction, computer science and network science help transform social signs and meanings into technological templates that organize reality. In this chapter, we look at how these sciences give shape to a network that structures sociality for its users, solidifying patterns, making some things possible and other things impossible, some things knowable or near, and other things unknowable or far.

Social Computing: Making People Usable

Phil Agre suggests that while information technology is said to be revolutionary, in reality it is quite frequently the opposite, as "the purpose of computing in actual organizational practices is often to conserve and even to rigidify existing institutional patterns."[4] If this is the case, we should ask which social patterns computing has solidified at the expense of which other ones.[5] In other words, in celebrating the hypersociability that digital networks open up to us, we would do well to keep in mind, as Paul Dourish reminds us, that "[o]ur experience using computers reflects a trade-off that was made fifty years ago or more."[6] The trade-off, to put it simply, is that the computer's needs are valorized over our own. This trade-off was not the result of secret agendas or ulterior motives, but one that emerged out of simple necessity: access to computers at the beginning of the cybernetic revolution was expensive; computer time was scarce and therefore more valuable than people time. This meant that interaction with computers had to be done in a way that made it *easy* for the computer, even if it made it difficult for us. We had to speak *its* language.

Since then, of course, things have changed to a certain degree. Computers are faster, cheaper, and smaller, which means they have moved out of the research lab first and into the home and the office; and now, they practically travel with us everywhere we go. Concurrently, there have also been significant improvements in trying to make it easier for us to interact with the computer in a more natural (or "human") way—to make the computer speak *our* language. Disciplines like human–computer interaction (HCI) are addressing this problem by trying to move away from designing *procedures* (a series of algorithms that perform assigned tasks) and toward analyzing *interactions*, the fluid interplay between machines and humans.

Nevertheless, we can basically look at the five major innovations in computing—high-level programming languages, real-time computing, time sharing, graphical user interfaces, and networking[7]—and see them not just as a history of the computer becoming better suited to humanity but as a history of the changes imposed on average individuals to make them better suited to the computer. Perhaps this sounds too much like the technological determinist joke about humans being merely technology's way of replicating itself. But under this alternate reading of the history of computing, we can see how these innovations (programming languages with more "natural" language commands and structures, graphical interfaces that lowered the barriers of entry so that masses of nonexperts could operate the machines, etc.) also make sense from the point of view of conserving those behavioral patterns that make the technologizing of the social—the conforming of humans to computers—much more effortless.

As forward-looking and revolutionary as concepts like *social computing* and *social media* may seem, the paradigm that established that we must accommodate the computer, and not the other way around, continues to influence the way we structure the integration of computers and humans. Social computing, after all, is approached by its practitioners as a way to use computer science to model, replicate, and predict social behaviors: "Social computing is an area of computer science at the intersection of computational systems and social behavior. Social computing facilitates behavioral modeling, which provides mechanisms for the reproduction of social behaviors and subsequent experimentation in various activities and environments."[8]

A critique of the premises behind social computing (the idea that computers can model complex social behaviors, describing a nodocentric world where we are able to model and predict the behavior of nodes) is not difficult to find, both outside and even inside the field. Such critiques are often reminiscent of Jaron Lanier's remarks about artificial intelligence (AI) in which he suggests that AI does not make computers smarter but people more stupid: "[P]eople are willing to bend over backwards and make themselves stupid in order to make an AI interface appear smart."[9] Likewise, social computing seems to invert the priorities, proposing that a reductive model of social behavior is more realistic than the real thing, and that humans do behave like computer programs. Giving voice to a sentiment that many Web 2.0 gurus would probably embrace, not realizing it is a critique, Trebor Scholz argues that the purpose of the sociable web (where many ideas from social computing have come to bear fruit) is basically to make people "easier to use."[10]

By virtue of the limits of computational models, the digital network does not facilitate all kinds of social behaviors equally, it merely conserves or solidifies those behaviors that can be observed, measured, and quantified. The implications that follow from the application of social computing and network science to social behavior need to be explored more closely.

Network Science and Network Scientism

If networks are indeed material structures (and not mere metaphors), it is still difficult to perceive or grasp them. This is because they are complex and distributed assemblages of things and people, spanning across multiple scales of time and space, which we are often not able to perceive with our direct senses. Therefore, it seems logical that we rely on the abstract reasoning of science to detect and measure networks. But network science does not merely describe networks. As Peter Monge and Noshir Contractor[11] state, it also provides the instructions for their design.

Network science can be defined as the organized study of networks based on the application of the scientific method. The scientific study of networks is not new. It began, arguably, with the branch of mathematics known as graph theory founded by Leonhard Euler in 1736. Since then, the principles of network science have been used to discover and describe relationships among everything from proteins to terrorists. Of course, the principles of network science have not remained static since the eighteenth century. More sophisticated tools for data collection and processing have translated into more complex network models. According to Albert-László Barabási,[12] science has recently contributed two important concepts to our understanding of networks. The first one is that the distribution of links in most networks found in the natural and social domains is not random but determined by logarithmic power laws, meaning that a few nodes (the ones acting as *hubs*, or central connectors) have many links, and conversely that many or most nodes have only a few links. This gives form to what is known as a "scale-free network," a network that can grow or expand easily. The second concept is that as these scale-free networks grow, they exhibit a form of "preferential attachment" whereby new nodes tend to link to older or bigger nodes, meaning that rich nodes get even richer in terms of the number of links connecting them to other nodes.

There are two other important concepts that science has contributed to our understanding of networks: "node fitness," an indicator of a node's ability to attract more links than others even if it has not been around for as long as older nodes in the network; and "network robustness," an index of how many nodes within the network would need to fail before the whole network would stop functioning altogether.[13]

There are also a variety of metrics that have been developed to quantify the behavior of networks. These metrics can describe the properties of the network as a whole, the behavior of individual nodes themselves, or the properties of the ties or links that connect the nodes.[14] For instance, properties of the network as a whole may include

- *size* and *density*, which indicate respectively the number of nodes in the network, and the ratio of actual links to possible links that could exist in the network;
- *centralization*, which measures the difference between the centrality score (a combination of closeness and betweenness—see further in the text) of the hubs and the rest of the nodes in the network;
- *transitivity*, which describes the degree in the network to which a triad of actors are connected in a close loop (whenever A is connected to B, B is connected to C, and A and C are also connected); and
- *inclusiveness*, which is a network metric that actually attempts to deal with the excluded by comparing the number of actors in the network to those not included in the network.

As far as metrics that describe the properties of nodes themselves, these may include

- *in and out degree*, which indicates the number of incoming or outgoing links to and from a node;
- *diversity*, or the number of links to nodes that have been classified as belonging to separate categories;
- *closeness*, or the average distance or degrees of separation of a particular node;
- *betweenness*, which measures the degree to which the node is in the path of one node to another; and
- *prestige*, or the degree to which a node receives links instead of being the source of outgoing links.

Properties that describe the links that connect nodes include

- *direction*, which indicates whether the link flows to and/or from the node;
- *indirect links*, which describe a connection that involves more than one degree of separation;
- *frequency*, which indicates how many times a link occurs;
- *stability*, or the endurance of a link over time;
- *multiplexity*, which describes more than one kind of link between two nodes;

- *symmetry* or *reciprocity*, which indicates whether a link is bidirectional; and
- *strength*, which describes the intensity of the link.

These metrics can be applied equally to the study of natural networks as well as social networks; one can speak, for instance, of the betweenness of a protein or a person to describe how it acts as an intermediary. Needless to say, this has changed the way we study and define social phenomena, insofar as people become nodes and social ties become links. Network science operates under the assumption that every social formation can be mapped and studied using the metrics described previously.

The branch of network science that uses networks as frameworks for understanding the structure of social systems is known as *social network analysis*. Throughout its seventy-year history, social network analysis has been used to study systems as small as families and as large as the world. Its goal has been to explain how the nodes in these networks make use of the links connecting them to exchange resources, ideas, and messages. In essence, social network analysis attempts to shed light on the mystery of how community is formed and maintained—a task made increasingly more complex by modern communication technologies, since they make it possible to establish communities no longer confined to one location in space.

According to Barry Wellman, technology has allowed communities to evolve from homogenous, "densely knit, geographically bounded groups" to "far-flung, loosely-bounded, sparsely-knit and fragmentary" groups.[15] Thanks to technology, individuals can move in a single moment between multiple, amorphous communities that occupy both local and global dimensions, and engage in interactions of varying intensity (from full engagement to a passing ambient awareness) with diverse peers. The study of these exchanges within networks is framed by Wellman in what he calls the two aspects of the "community question": "How does the structure of large-scale social systems affect the composition, structure, and contents of interpersonal ties within them?" and "How does the nature of community networks affect the nature of large-scale social systems in which they are embedded?"[16] In other words, how does the network influence the node, and how do the nodes influence the network? This resonates with Van Dijk's[17] observation about the double affordances of networks, which can facilitate two processes at once: the algorithms of the network can influence the social behavior of the users at a macro level, while at the same time the aggregate of interpersonal exchanges *become* the social content of the networks. Social network analysis can help us map this dynamic as we try to answer the community question.

However, we must remain critical of the way these scientific concepts are applied when it comes to the modeling of networks as templates for certain kinds of sociality.

There are two main concerns in the application of network science to the study of digital networks: the assumption of scarcity as the determining factor in interaction, and the constraining of research questions by the available metrics for the study of networks. I will outline each one briefly.

Social network analysis, whose history obviously predates digital networks, has always assumed a scarcity of resources in society. Thus social network analysis focuses on the "structural integration of a social system and the interpersonal means by which members of this social system have access to scarce resources."[18] One of the concepts in social network analysis that attempts to explain the importance of ties or links to overcome scarcity is the concept of social capital.[19] Nodes with more links (more social capital) are believed to have a greater chance of overcoming scarcity. Not surprisingly, a lot of effort has been expended in figuring out how to design social networks where nodes can conduct the transaction of social capital favorably. For instance, Monge and Contractor[20] have identified eight simple rules of communication that govern the exchange of social capital in all social networks. The rules are as follows (each rule is based on scientific theories that are beyond the present goal to describe, so they have just been listed for reference): nodes try to keep the cost of communication at a minimum (theory of self-interest); nodes try to maximize the collective value of their communication (theory of collective action); nodes try to maintain balanced interactions among those they communicate with (balance theory); nodes are more likely to communicate with someone who has what they need or need what they have (resource dependency theory); nodes are more likely to communicate in order to reciprocate for past exchanges (exchange theory); nodes are more likely to communicate with others who are similar and not with others who are different (theories of homophily); nodes are more likely to communicate with others who are physically near or electronically accessible (theories of proximity); and nodes are more likely to communicate with others in order to improve their individual fitness or the fitness of the network (coevolutionary theories).

Directly or indirectly, these rules to overcome scarcity have been incorporated into the design of digital networks. Through the algorithms of social computing, an image is presented of a *homo economicus* determined to manage this scarcity. But these rules also restrict our understanding of actors in a network: in its quest to overcome the perceived scarcity through the management and control of flows, social network analysis reduces sociality to a set of prescribed network relations. In other words, the design of digital networks has taken these scientifically derived *descriptive* observations of behavior in networks and, by programming them into the code that regulates interaction among nodes, transformed them into *normative* rules of behavior. That this translation from descriptions to rules has taken place is not surprising, since—to a certain extent—the application of scientific knowledge in the creation of systems is what technology, as a practice, is all about. What should be open to critique, however, is the deployment of these rules in such a way that they become tools of domination, presenting obstacles to the creation of alternative forms of social organization.

While this kind of critique is the object of the second part of the book, what should be made clear at this point is that a critique of networks needs to transcend the boundaries of a network epistemology. In short, we must ensure that the questions we ask about networks are not subordinated to the solutions network science can provide. Part of the problem is that the practice of science as an exercise through which the measurable properties of nature or society are revealed has many advantages but at least one frequent disadvantage: focus tends to be placed on formulating research questions that can be answered with the models, laws, and theories already at our disposal instead of developing new questions whose solutions might not be as readily available.[21]

In describing nodes, links, and networks in terms of specific metrics (betweenness, transitivity, etc.), or analyzing behavior in terms of particular communication theories (self-interest, balance, etc.), we might neglect to consider other important dimensions to the study of the digital network that might not be as easy to quantify or measure (such as questions about the degree to which participation increases inequality, or questions about the degree to which the outside of nodes represents an ethical resistance to network logic). This contributes to what Manuel DeLanda describes as the illusion promoted by scientism: "[T]hat the actual [measurable] world is all that must be explained."[22] Nodocentrism is thus a form of scientism, a belief that only nodes are real and only nodes deserve to be explained, and that we need only quantifiable measurements to describe and predict their behavior. Questions about what alternative ways of looking at digital networks might look like are silenced before they can be asked, because we are only interested in solutions that can be measured with the metrics and rules network science has identified. The process through which alternatives are generated is irrevocably arrested.

The reasons why there is such an investment in network science as the study of unchanging principles used to build templates for organizing society are not difficult to discern. A report on network science commissioned by the U.S. military states, "Network science consists of the study of network representations of physical, biological, and social phenomena leading to *predictive* [my emphasis] models of these phenomena."[23] This predictability is necessary because networks are seen as having weaknesses that can be easily exploited by disruptive forces. The same report goes on to explain that these disruptions can only be addressed through superior network design: "Large infrastructure networks evolve over time; society becomes more dependent on their proper functioning; disruptive elements learn to exploit them; and society is faced with challenges, never envisaged initially, to the control and robustness of these networks. Society responds by adapting the network to the disruptive elements, but the adaptations generally are not totally satisfactory. This produces a demand for better knowledge of the design and operation of both the infrastructure networks themselves and the social networks that exploit them."[24]

In short, we are in a race to build better and more resistant networks before they become overrun by disruptive elements such as terrorists or hacker collectives. As what might be expected, the race to design and control these improved digital networks starts with the algorithms.

Social Allegories and Algorithms

Network science, social network analysis, and social computing provide the frameworks not only for understanding but also for building the complex assemblages that are digital networks, the assemblages that in turn act as determinants of social behavior. To better understand how these frameworks are actually codified into the architecture of digital networks, I will attempt to establish a link between the computer algorithms of digital networks and the social allegories contained in them.

First, a word about allegories: traditionally, we think of allegories as literary or artistic devices that are used to convey a meaning that is other than the literal meaning. Thus we can think of works like Fritz Lang's *Metropolis* or William Golding's *Lord of the Flies* as allegories (their surface narratives hint at deeper insights about technology, human nature, etc.). Here, however, I will be using the concept of the allegory more loosely, simply to imply a device that is used to transfer meaning through symbolism from one context to another. In essence, I will be arguing that computer algorithms can communicate meaning through allegories between the realm of social behavior and the realm of network architecture. My exploration of algorithms and allegories in digital networks is motivated by two questions: If *algorithms* are formulas or processes for solving problems, what are the social problems that the algorithms of digital networks intend to solve? And if *allegories* communicate meaning through symbolism, do the algorithms of digital networks function as allegories that convey a message about the social in the act of "solving" these problems?

Digital networks are computer programs, so they contain algorithms. These algorithms transform user actions (like clicking a "Like" button in Facebook) into a series of predefined operations in the network (increasing the number of likes of the digital object in question by one, adding the object to the list of things the user likes, and so on). But in doing so, they assign a slightly different meaning to what it means to "like" something (or to "friend," "chat," "recommend," "join," and so on) in the digital network. This meaning, however, is not entirely new. It is based on common understandings of what it means to like, friend, chat, recommend, or join something *outside* the network, in the so-called real world. This way, the algorithm serves as an allegory of sorts by establishing a correspondence between two operations—albeit with the same name—in two different realms of social meaning (e.g., to like something in the digital network, and to like something in "real life"). To "friend" someone in a social networking site therefore implicates an algorithm that references the social act of forming a friendship in an allegorical way and codifies that act as a set of computer processes (establishing a correspondence between two records in a dataset, for instance). In this manner, network metrics (such as the centrality of the friend, the frequency with which the friend links to us, etc.) not only are used to construct algorithms that serve as allegories of social acts but also redefine or give new meaning to those acts in the process.

Much like a video game player discovering the meaning of certain actions in the game, and discovering which sequence of actions has what set of consequences, the digital network user learns to play the algorithms of the digital network. A digital network, like a video game, is a complex system of meaning that assigns a quantifiable value to the elements within it, which therefore establishes an economy that can be discovered through participation. By playing the algorithms of the network, the user discerns the mechanics of the economy (e.g., how acquiring more friends, gaining more incoming links, or contributing more content might result in more visibility within the system).

To repeat a point made earlier, the design of digital networks takes scientifically derived observations of social behaviors and, by converting them into computer code that regulates interaction, transforms these observations into algorithms that facilitate certain forms of action (and obstruct others). In the process, social acts are given new meaning, although they continue to allegorically reference the original act outside the network. Thus to talk about a "recommendation" in the realm of digital networks is to reference the act of one person suggesting an object (like a book or movie) they think another person might enjoy. But within the network system, a recommendation is nothing more than the application of algorithms like *collaborative filtering, naïve Bayes classifiers, decision-tree classifiers,* or *k-nearest neighbors*[25] to calculate the probability that a user will be interested in a particular object. The algorithmically derived recommendation is only an allegory of an interpersonal recommendation because it is made by a machine, not a human (even if the machine is merely aggregating lots of human opinions); it is derived from preferences the user has disclosed to the network, not from personal knowledge. Insofar as they allegorically stand for human interactions, these computer operations are only possible to the extent that we allow our behaviors to become legible to the algorithm.

The algorithms of digital networks enforce certain modes of social conditioning. Nodocentrism means that things not rendered as nodes are practically unintelligible to the network, which suggests that being *in* the network requires nodes to continuously participate in ways that makes their behavior legible to other nodes, in alignment with network logic. Although much has been said on how decentralized networks spell the end of censorship, we are only just beginning to understand how participation in networks fosters certain kinds of self-censorship: we have to learn which behaviors we want to highlight so that they can be seen by the network and which ones we want to avoid so that they cannot be misinterpreted by the algorithm.[26] To behave in a way that does not conform to the logic of the network means to render oneself invisible, to cease to exist. The economics of the network are such that a node's existence depends on its ability to obtain attention from others, to allow its movements to be monitored and its history to be known.

The Agency of Code

While algorithms will probably continue to afford increasingly sophisticated social operations, it is important to realize that a lot of what we now consider the social revolution that is the Internet has occurred as a result of connecting computers and humans in relatively simple ways and letting complexity emerge out of the aggregation of lots of simple social operations. As a way of providing a brief illustration of how the code of digital networks can assume social agency and control in this simple manner, I will discuss the example of a type of Web 2.0 application called a social tagging system.

A social tagging system allows a network of users to classify resources by assigning descriptive tags or keywords to them (e.g., if I upload or encounter a picture of a cat, I might want to tag it with the keyword "cat" or any other keywords that I choose). Some of the most popular social tagging systems include social bookmarking applications like Delicious .com and photograph annotation sites like Flickr.com. The most important feature of social tagging system is that they do not impose a rigid classification scheme. Instead, they allow users to assign whatever classifiers they choose. Although this might sound counterproductive to the ultimate goal of a classification scheme, in practice it seems to work rather well. There is no authority—human or algorithmic—passing judgment on the appropriateness or validity of tags, because tags have to make sense first and foremost to the individual who assigns and uses them. While tags serve primarily a personal purpose, facilitating the retrieval of resources by the individual at a later time, the use of the same tag by more than one person engenders a collective classification scheme known as *folksonomy* (a portmanteau of *folk* and *taxonomy*). The whole point of a social tagging system is that the aggregation of inherently *private* goods (tags and what they describe) has *public* value: when people use the same tag to point to different resources they are organizing knowledge in a folksonomy that makes sense to them and others like them. In other words, the tag is the object that brings a resource and a social group together via the shared meaning of a word.

We can say, then, that the social tagging system functions at the intersection of individual choices and the shared linguistic and semantic norms of a social group (the *folks* in folksonomy). The code of social tagging systems may not directly force users to employ certain kinds of tags, but by indirectly raising the expectation that tags might be useful to others, it shapes social activity in the process of aggregating individual tagging choices into collective information.

The Delegation of Meaning

The greatest strength of social tagging systems is also perhaps their greatest weakness: the way in which the negotiation of meaning during the process of classification is delegated from humans to code. Decisions regarding how to classify things, which used to be undertaken by humans in collectivity are now carried out by humans individually, while the code aggregates and represents those decisions. If we see this as a replacement for oppressive systems of classification in which one group of people used to impose their classification scheme on the rest, this might be seen as an improvement. If we see this as a replacement for democratic systems in which equals used to negotiate and collaborate on the definition of a classification scheme (and in the process gave shape to what defined them as a group), then the outcome might not be as positive. This is because this process is now conducted by the code, without some of the opportunities for negotiation and collaboration that other paradigms afford. As is always the case with technology, where the line is drawn between the open affordances of social tagging systems (what they facilitate and what they constrain) depends on how the technology is applied.

In order to understand how code assumes social agency in social tagging systems, we must first contextualize the manner of classification that these systems embody. There are two ways in which a classification system allows for meaning construction. One is in the use of the system to search for resources already in the system. The other is in the contribution of new resources to the system. A traditional classification system, based on a structured taxonomy, guides users in search of resources by moving from the general to the specific, at each branch presenting clearly defined options. Imagine you wish to find a resource using the

Yahoo! Directory. Does the resource have to do with arts and humanities, business and economy, or one of the other categories? If it is related to arts and humanities, does it have to do with photography, history, literature, or one of the other categories? Yahoo! decides what those categories are, and individuals use their familiarity with the classification structure to find things. Now imagine you wish to add a resource to the system. In that case, you would use the same categories to find the appropriate place for the resource. If such a category does not exist, then the administrators of the system must decide whether it needs to be created, and where in the overall scheme it needs to be added.

Folksonomies differ from this structured taxonomy approach in significant ways. The most obvious one is that any user of the system can create tags or categories without permission from any kind of authority. Another important difference is that tags need *not* be arranged in any particular way. If the tag or category *cat* is close to the tag or category *car* it is because of alphabetical reasons, and not because the proximity of *cat* and *car* says something about any of the two signified elements. Because categories do not occupy a specific location in a structure, folksonomies allow for the association of an infinite number of tags to a resource. In other words, a picture of a cat driving a car can be marked with both tags and included in both sets, as well as any others that the user chooses.

Another difference between folksonomies and structured taxonomies that might not be so obvious is the role of human collaboration in their definition. Structured taxonomies require consensus in the form of at least two collaborating human subjects (whether this consensus is achieved democratically or hegemonically is another topic). If a category is defined but no one adheres to it, can it be said to exist? Folksonomies, on the other hand, do not require consensus as much as they measure the consensus already established around the use of certain words. In other words, folksonomies assume consensus without involving humans in the process. Social tagging system users have no discussion whatsoever about how categories should be defined, what they mean, or their relation to each other. Instead, what the code cares about is that if two people used the tag *cat*, it will aggregate and display the resources associated with that tag, regardless of whether one user meant the furry feline and another the Center for Alternative Technology. Of course, if the latter user had employed that tag *CAT* instead of *cat*, the code would react differently (which perhaps means, as Clay Shirky suggests, that there are no such things as synonyms in a folksonomy[27]).

In essence, the code of social tagging systems removes the need for humans to negotiate meaning around classification. This can be liberating as well as alienating: it is liberating because, as I suggested earlier, there is no governing body dictating what the classification scheme should be; and alienating because, without the mechanisms for deliberation, meaning becomes atomistic, a reflection of what the software has parsed and aggregated from detached individuals, not what has emerged through consensus and deliberation.

By this I do not mean to imply that social tagging systems do not open up all kinds of new social operations heretofore impossible (they are, after all, *social* media). I merely want to call attention to this different way in which we are defining and constructing sociality—a sociality that is the result of code doing things to the resources of detached individuals. There are plenty of social transactions that can be carried out in social tagging systems, such as being able to see different items classified by different people with the same tag, or the same item classified by different people with different tags, or the resources of a particular individual, and so on. But the scope of these affordances is defined by the code, and the community willingly relinquishes a large part of its agency in exchange for individual freedom and the scale of access that only the Internet can provide.

While the benefits of this freedom and scale are obvious, some people rightfully point out the risks of surrendering agency in the process of negotiating how knowledge should be structured. Shirky, representing arguments focusing on freedom and scale, states in reference to Delicious.com that "aggregate self-interest creates shared value. . . . By forcing a less onerous choice between personal and shared vocabularies, del.icio.us shows us a way to get categorization that is low-cost enough to be able to operate at internet scale, while ensuring that the emergent consensus view does not have to be pushed onto any given participant."[28]

On the other hand, Matt Locke describes the functions relinquished by the community and how the code assumes those functions in some form or another: "There are no politics in folksonomies, as there is no meta-level within the system that allows tagging communities to discuss the appropriateness or not of their emergent taxonomies. There is only the act of tagging, and the cumulative, amplified product of those tags."[29]

It is in discussing this "appropriateness" that social groups in fact define themselves. Clearly, there *are* politics in folksonomies, but we need to uncover them by asking not only what kind of social agency the code assumes on behalf of the networked subject but also how this conforms the networked subject itself.

4 ACTING INSIDE AND OUTSIDE THE NETWORK

DIGITAL NETWORKS MEDIATE our social realities according to templates where certain forms of sociality are algorithmically operable and others are impossible for the algorithm to perform. Because these templates are increasingly subordinated to for-profit interests, it is important to explore how they structure the formation of the self, and what other models for conforming the self outside these templates are available. However, the problem with framing the question this way is that it already presupposes a separation between our networked and unnetworked selves. A neat separation between a networked world and a world that remains untouched by digital networks is increasingly difficult to maintain, even for the purposes of conducting a critical analysis. If half of the population of the planet has a cell phone, it is nearly impossible to talk about dimensions of life not affected in a direct or indirect manner by the network as it mediates or governs the relationship between the individual and the social. Thus to theorize the networked subject is also to theorize the ways in which the digital network has become a universalizing logic for ordering the social and providing certain types of agency.

Until very recently, it was convenient for those wishing to engage in a critique of the network to establish a false dichotomy between our networked selves and those parts of our lives not connected to digital networks. Because the Internet and other digital networks were still new, and not yet such a pervasive presence in our lives, we developed a convenient habit of seeing our online experiences as unfolding in an alternate part of reality. We believed that our actions could begin, unfold, and conclude entirely online without any repercussions to life offline, thus concluding that virtuality had its own set of rules and values that did not correspond in a one-to-one manner to the rest of reality. But even during this early period, many critics began to look at the rupture between the two realms, and the possibilities supposedly afforded by the virtual domain, with skepticism. Albert Borgmann, for instance, compared online and offline communities and suggested that commodification (to take something that is outside of the market and inscribe it in the market) was the distinguishing feature that separated the former from the latter. Borgmann argued that online communication itself reduced everything to an economic exchange meant to secure attention from others: "The internet is culturally commodifying by its nature. . . . What happens in fact is that commodification reduces ourselves and those we encounter on the internet to glamorous and attractive personae. Commodification becomes self-commodification, but shorn of context, engagement and obligation, of our achievements and failures, of our friends and enemies, of all the features that time has engraved on our faces and bodies—without all that we lack gravity and density."[1]

In contrast to these commodified communities, Borgmann described what he called "final communities": "[F]inal communities are ends rather than means, or more precisely, they are the groups of people where one finds or works out one's reason for living. . . . The point is that final communities require the fullness of reality, the bodily presence of persons, and the commanding presence of things. Any attempt to secure the fulfillment of one's deepest capacities and aspirations in and through cyberspace will founder on the shoals of commodification."[2]

Consequently, Borgmann saw any attempt to form final communities by using the Internet as bound to fail: "Use of the internet at home leaves people feeling lonely and unhappy."[3]

A similar critique is posed by Hubert Dreyfus who, following Kierkegaard, argues that to escape the anomie of modernity one needs to form unconditional commitments. This type of commitment establishes "qualitative distinctions between what is important and what is trivial, relevant and irrelevant, serious and playful"[4] in life, determining what we hold to be significant in it. Unconditional commitments make us vulnerable, because what we hold to be true may disappear or turn out to be false. But it is precisely this risk, according to Kierkegaard, that produces a strong identity and gives individuals a perspective on the world. Dreyfus then wonders whether the Internet can encourage and support unconditional commitments. He concludes that, similarly to Kierkegaard's assessment of the press and the public sphere, the Internet does not necessarily prohibit but definitely *undermines* unconditional commitment:

"Like a simulator, the Net manages to capture everything but the risk. . . . [I]f we are sufficiently involved to feel as if we are taking risks, the simulations can help us acquire skills. But insofar as [these simulations] work by temporarily capturing our imaginations in limited domains, they cannot simulate serious commitments in the real world . . . the risks are only imaginary and have no long-term consequences. The temptation is to live in a world of stimulating images and simulated commitments and thus to lead a simulated life."[5]

Dreyfus ends by arguing that unconditional commitments can only be formed when the identities, knowledge, and skills we develop online are transferred to the real world, where the risk becomes real. This is, however, practically impossible according to Dreyfus because the nature of online experiences inhibits this very step: "Indeed, anyone using the Net who was led to risk his or her real identity in the real world would have to act against the grain of what attracted him or her to the Net in the first place."[6]

In retrospect, these kinds of arguments essentialize—to the point of oversimplification, perhaps—the online and offline worlds as two distinct realms of reality, with no intersections between the two social realms. What undermines them is that they establish a very clear separation between the self as it exists within the network and the self as it exists outside it, in some sort of a "natural" social order that is corrupted or complicated by the arrival of digital network technologies. Merely a few years later, this kind of critique sounds quaintly absolutist. Because of the ubiquitousness of the digital network, it is possible for our networked existence to encompass all the dimensions of our social lives. In other words, it has become not only possible but also commonplace to extend our interactions with our most intimate acquaintances through the digital network. Even digital natives (those generations exposed to digital network technologies from birth) will admit that online experiences are indeed no substitute for the "real" thing; however, the point—they will add—is not to replace the "real" thing, but to supplement or augment it.

Thus the digital network has not done away the real. It has merely converged with it. Mediated social exchanges have become so intertwined with unmediated ones that it is no longer possible (or necessary?) to tell where the real and the simulated begins or ends. Something can start as an exchange on an electronic forum, move to a private face-to-face conversation, continue over text messaging, and so on. Hence the futility of talking about the networked subject as if it was an avatar, a member of a parallel community whose actions concern a separate universe (the digital network). Rather, to talk about the networked subject is to talk about a fragmented self, some of whose multiple identities are wired or connected to the network and some that are not. Nonetheless, there is no way to understand this fragmented self without appreciating how network logic mediates the perception of reality for the subject, how it constructs the agency models for the subject to act in concert with technology, and how it establishes a new form of social contract to replace the model of subjecthood previously granted by other social institutions.

Mediating the Networked Self

How do digital technologies intervene to mediate the world of the individual? Are being and knowing, as mediated by the digital network, qualitatively different or even inferior than those forms of being and knowing that are not mediated by the digital network?

The false dichotomy between the networked or mediated self and the unnetworked or unmediated self mentioned earlier seems to have distant echoes in tropes such as Plato's allegory of the cave, where we encounter the idea that what we perceive as reality is an illusion, and that the authentic (unmediated) version of reality is *out there* waiting to be grasped by those minds capable of true understanding and learning. This has been a common theme in Western thought, and in more contemporary times, works of fiction like *The Matrix* have hinted at the role that technology can play in making the *illusion* of reality more realistic and, thus, more pernicious. While truly immersive virtual reality (undistinguishable from reality) remains a fantasy, modern communication technologies have succeeded in producing a disembodied subject that can experience alternative or enhanced forms of reality. In other words, by allowing us to know or experience the world indirectly, technology can put us in places without having to be physically there. This technologically mediated sense of detachment from local space and reattachment to hyperspace is known as "telepresence" or the experience of being somewhere where our bodies are not. Telepresence has become a routine experience for most of us, as common as talking to someone on the phone. Through telepresence, as Dreyfus says, "our bodies seem irrelevant. . . . our minds seem to expand to all corners of the universe."[7] But when our interaction with the world is reduced to mediated signals, how do we know if things on the other side are real? How do we assign to them the appropriate importance?

These questions have preoccupied philosophers well before the advent of modern communication technologies, of course. Descartes, for instance, was concerned not with the reality of things on the other side of the screen, but on the other side of the brain. He believed that all we have access to in the world is our private experience. The world, in his opinion, was out to fool the brain, the only reliable organ through which we could assess the reality of things. Skeptics of the "realness" of reality had been around before him, but Descartes was really the first one to question the reality of perception. He did this on the grounds that the sense organs (the eyes, ears, nerves, etc.) are unreliable transmitters of information to the brain, which is the only one capable of interpreting and acting on that information. According to his model, our access to reality is indirect, mediated by the senses but actualized exclusively by the brain. This line of thinking lead Descartes to believe that the only thing we could be certain of was the content of our brains, and everything in the outside world was consequently less real or not real at all. This skepticism about the existence of the external world actually fueled the development of the branch of philosophy we know as epistemology, which concerned itself with assessing the validity of our everyday beliefs about the world.

While Cartesian epistemology was gradually replaced with other approaches for making sense of the world that do not presuppose a separation of the mind and the external world,[8] Dreyfus suggests that because digital network technologies are making perception more and more indirect, and demanding that we take for granted the reality of what we perceive, Descartes's epistemological doubts are being resurrected. As a result, it sometimes seems as if the networked subject occupies a Cartesian plane where the only thing that can be taken for granted is the self, and every other aspect of networked reality is a world of mediated shadows whose reality we can only infer.

Biases in the Way the Network Mediates Involvement

While we might not want to go as far as questioning the reality of every single thing mediated by the network, it is necessary to at least be cognizant of the ways in which reality is processed by the network. As digital networks mediate social reality, they tend to favor certain types of involvement over others, which constitutes the networked self in one manner and not in others. What I describe next are trends in how social involvement is structured in the network. This is not meant to imply that every single digital network exhibits all these characteristics all the time. Rather, what is meant is that as networks operate according to the principles of nodocentrism, their architecture seems to predominantly (although not always exclusively) exhibit a bias toward the following processes.

Immediacy. Immediacy usually indicates the distance across space between social actors, and is thus a factor that can impact our sense of social involvement. In digital networks, however, *spatial* distance no longer seems a relevant metric given that—we are told—technology annihilates distance. Thus immediacy becomes a function of those metrics in the network that express closeness, regardless of a node's geographic position. The digital network exhibits a bias toward expressing immediacy or nearness in nodocentric terms. This does not mean that the network obstructs a sense of immediacy, but that it exhibits a bias toward presenting that which is networked as near, whereas that which is not networked is perceived as far.

Intensity. Intensity describes the strength with which actors perceive social acts to the exclusion of other phenomena. For example, face-to-face conversations have high intensity because of the amount of information coming from one source, whereas an online text chat comparatively has lower intensity. This does not necessarily mean that high intensity social scenarios are "better" than low intensity ones on some sort of moral scale. Nor does it mean that a series of ongoing low intensity interactions cannot feel quite "intense" (low intensity allows for multitasking, since the user can quickly switch between a number of simultaneous exchanges). It just means that intensity in this context can help us understand how networks redistribute an individual's attention and energy across networked sites. Digital networks are biased toward low-level intensity social interactions because this kind of involvement is more cost-effective and less time consuming than high-intensity interactions.

Intimacy. Borrowing from Joseph B. Walther,[9] we can describe the intimacy of social interactions according to three categories: impersonal, interpersonal, and hyperpersonal. Impersonal interactions are those that contain low levels of personal information, and they are ideal for situations where a goal can be accomplished without a significant exchange of information about the participants. Interpersonal interactions are those that contain higher levels of personal information and allow participants to develop or sustain social relationships beyond just getting the job done. And hyperpersonal interactions are those in which actors have technological means to control the personal information they wish to share (means that they would not have available in regular face-to-face interactions). Hyperpersonal social interactions thus involve the ability to form interpersonal social relations "without the interference of environmental reality."[10] Digital networks have a bias toward supporting impersonal and hyperpersonal social involvement. Privacy settings, for instance, are a tool of hyperpersonal social involvement because they allow the participant to decide which aspects of their personal information to share, or not to share, with a degree of control that would not be possible outside of the digital network (the fact that many users have not yet realized the consequences of setting these privacy settings correctly is a separate issue).

Simultaneity. Asynchronous forms of communication allow us to communicate without having to be concurrently engaged with the person we are exchanging messages with. And while electronic media are sometimes associated with the advent of a second age of orality, asynchronous communication—from e-mail to text messaging—continues to be a major feature of digital networks.[11] Thus digital networks have a bias toward nonsimultaneous social involvement.

There are the trade-offs to this loss of simultaneity. Schutz defines simultaneity as the ability to experience our consciousness in parallel with another human being's consciousness through the act of communication or interaction. He writes, "[W]hereas I can observe my own lived experiences only after they are over and done with, I can observe yours as they actually take place. This in turn implies that you and I are in a specific sense 'simultaneous,' that we 'coexist,' that our respective stream of consciousness intersect."[12]

The outcome of this experiencing of parallel subjectivity is not that we are able to read each other's minds. It is simply the realization that one is experiencing a fellow human being (which is, I suppose, what the Turing test seeks to replicate). To be sure, simultaneity can be approximated through other forms of mediated interaction; some digital network technologies (like voice or video chats) can support real-time or synchronous communication. But as a general rule, we can say that the larger the network, the more pressing the need for efficiency through asynchronous management of communication among the participants, which means the fewer the opportunities for members to experience simultaneity.

Simultaneity is time consuming. Digital networks might make it possible for more people to be on the network at the same time, but as the number of links or "friendships" increases, the possibility of having a truly simultaneous intersection of streams of consciousness with most of those people decreases. Since information in the network must circulate at ever-increasing speed and efficiency, social interactions become predominantly nonsimultaneous.

Because of the reconfiguration of immediacy, intensity, intimacy, and simultaneity, we could say that the digital network exhibits an overall bias toward engagement with contemporaries as opposed to consociates. According to Schutz, consociates are the individuals I can experience through simultaneity. Contemporaries, on the other hand, are the people I know exist but whose consciousness I cannot experience in parallel. Schutz says of the latter, "[W]hile living among them, I do not directly and immediately grasp their subjective experiences but instead infer, on the basis of indirect evidence, the typical subjective experience they must be having."[13]

The necessity to manage time and resources means that social interactions in digital networks tend to become identified with disembodied immediacy, low intensity, guarded intimacy, and nonsimultaneity. As we saw in the example of social tagging systems, the subjectivity of network users can only be inferred "on the basis of indirect evidence" (such as tags) through the manipulation of digital objects. Thus although networks can facilitate interaction between consociates through mediated synchronous interaction, they have a bias toward mediating social realities where interaction between people (contemporaries) is increasingly supported nonsimultaneously. This is not to say that to a digital native, communication via these technologies can indeed feel immediate, intense, intimate, and simultaneous. But the point of the previous discussion was to frame how these biases are predominant in the network, and what impact they have on the self.

Mediation and the Obstruction of Being

Even though I have warned against the danger of talking about the networked self as if it was a separate self with a separate reality, perhaps I have myself promoted this contradictory approach by talking at various points about a *networked* subject. I have engaged in this practice merely with the intent to identify certain characteristics and critique them, and I will now continue to do so in order to question whether the digital network's mediation of the self is in some way obstructing the process of *being* in a philosophical sense. Again, the goal is not to reify a separate self, but to suggest that, if our networked and unnetworked selves are inexorably linked, there is no way to talk about the obstruction of being in one instance without considering the repercussions for the other. In essence, the problem is that prior to digital networks, we never had such a dominant or propagated model of technologically facilitated mediation of reality, one that left little room for alternatives. So the question of whether the digital network obstructs being is particularly pressing, even though the same question could (and should) be posed in regard to other technologies.

According to Theodore Rivers,[14] technology in general subverts being by demanding that our attention and efforts be placed at its service and by reducing the amount of time and effort we dedicate to things like contemplation and reflection. This is because, Rivers argues, technology is concerned with a reductive, repetitive form of action or endless doing: "Technology inhibits deep thinking because it is concerned primarily with activity, not contemplation. Because thinking is fundamental to self-awareness, technology is an obstacle to self-identity. It is a threat to internality."[15] Contrary to a view that sees actions as emanating from being, technology promotes an interpretation of being as emanating from action: *I do, therefore I am.* By virtue of what it was designed to do, technology fulfills its mission only as long as we are engaged in doing things with it; it occupies the self with continuous action and is unconcerned with what kind of being results from that action. Whereas Martin Heidegger once saw the premodern engagement with technology as a way of revealing the truth of being in the world ("There was a time when the bringing-forth of the true into the beautiful was called techne"[16]), now, as Rivers puts it, "[t]he relationship has been reversed: that is, technology is no longer an aid in the perfection of being, but rather being is now an aid to the perfection of technology."[17]

The worst reading of this argument would present us with a state of affairs in which technology has total control, and we are merely its puppets. However, this view would also assume that the networked self does not have any agency at all. Which brings us to the question of who or what exactly has the power to act in digital networks.

While it is true that digital networks shape our perceptions of social reality, it is also true that we can actively intervene in the shaping of those realities. Agency in digital networks—the opportunities to shape and transform those networks—is shared by humans and technology. When we intervene as technology's designers or masters, we can responsibly delegate agency to it, and allow it to perform certain social functions on our behalf (sometimes, to be sure, with unexpected consequences). However, when we become technology's subjects, we irresponsibly or involuntarily surrender our agency and allow technology to act, perhaps even against our interests. Therefore, the issue of how agency comes to be delegated is extremely important.

While modern history has positioned humans as masters and technology as the servant or slave (merely a tool to exercise our mastery over nature), dystopian critiques have presented a very different picture, with humans as the slaves of an autonomous technological master. But can technology act on its own? There are basically three theoretical approaches to allocating agency between humans and technology: realism, social constructivism, and what Philip Brey (2005) calls hybrid constructivism. Each approach attempts to answer in its own way the question of *who* acts in technosocial systems.

Realism, also known as technological determinism, establishes that technology shapes individuals and society. Technology prescribes behaviors and determines social practices. Think, for example, of a traffic light. It is nothing more than a mere flashing red light. Yet we obey it unconditionally (most of us, anyway) and organize our behavior around it. This is a simplistic example, of course, but while technological determinism might not go as far as ascribing consciousness or intelligence to technology, it does grant it the ultimate power to shape our social environments. Thus agency is a principal attribute of technology in this perspective, to the extent of treating artifacts as autonomous agents. Under this view, technology is definitely the master.

In the social constructivist perspective, it is we who shape technology: society's behavior and practices give technology its meaning. Agency cannot be attributed to artifacts, because the supposed "acts" of technology can always be traced to the actions and interpretations of social groups. *We* invented the traffic light; *we* came up with the system of laws, regulations, and technologies into which the traffic light fits as a constituent of a complex social system, and *we* can change any of that at any point. A traffic light in a deserted intersection has no purpose. Under this view, we are the master, and technology the servant.

These two approaches deliver us into a conceptual paradox. Which comes first: technology that creates social circumstances or social circumstances that give shape to particular technologies? As a way out of that paradox, most philosophers of technology have abandoned the master–slave dialectic altogether and aligned themselves with a third approach: hybrid constructivism.[18] Hybrid constructivism avoids making a discreet distinction between society and technology when it comes to the ability and opportunity to act. It suggests that technologies possess the potential to act, but this potential is only realized when they interact with other elements in social assemblages. This is a crucial point: the potentiality of artifacts (or humans, for that matter) is actualized only when they are part of a network of human and nonhuman actors. In hybrid constructivism, there are no clear-cut masters or slaves, since it is not possible to apportion agency exclusively or neatly to one party given the dependencies and interactions created through network transactions. In short, actors acquire their agency only as nodes in a network. Agency cannot exist in a social vacuum; without each other, human and technological actors cannot actualize their agency: "Agency is not, to be somewhat precipitous, rooted in the properties of entities-in-themselves, but rather in the properties of entities as elements of networks (or structures). And those networks/structures are invariably concatenations of both human and nonhuman actors."[19]

This hybrid approach is similar to another well-known theory of agency: actor–network theory, or ANT.[20] ANT establishes that "[a]rtifacts and their properties emerge as the result of being embedded in a network of human and nonhuman entities. It is in this context that they gain an identity and that properties can be attributed to them."[21] Since everything—human or technological—can be an actor on equal terms, hybrid constructivism in general, and ANT in particular, introduces a *generalized symmetry* in accounting for agency within a network: "The term 'hybrid constructivism' can be taken to refer to any position that adopts the principle of generalized symmetry. This is a methodological principle according to which any relevant elements referred to in an analysis (whether 'social,' 'natural,' or 'technical') should be assigned a similar explanatory role and should be analyzed by the same (i.e., symmetrical) type of vocabulary."[22]

This tendency to see all actors on equal or symmetrical terms is not without its problems. First, it might fail to provide meaningful or complex accounts of new social formations. Paradoxically, what is supposedly an empirical attempt to describe things *as they are* ends up obscuring any explanation of how the nodes came into being in the first place. By simply calling any assemblage a "network," social theorists end up confusing, according to Bruno Latour, "what they should explain with the explanation"; they begin with networks as self-evident explanations, "whereas one should end with them"[23] after explaining how they are constituted. In other words, the mapping of the network serves as the starting and end point; the distribution of agency is traced, but not explained, or explained by blackboxing motivations under the name of various "social forces." Or as J. Macgregor Wise puts it, "Agency cannot be so unquestioned. How do we account for differences (even similarities) in agency, in the distribution of agency? And how do we do this without recourse to abstract macro-actors such as social forces . . . ?"[24] The purpose of ANT is not simply to draw a network, but to try to explain the associations formed within it without resorting to these black boxes or abstract terms like "social forces."

Second, the tendency to see all actors on equal terms might obfuscate the special nature of human agency and, more important, human responsibility in technosocial systems. While the actions of technologies can be predictable to an extent (their affordances are materially circumscribed), hybrid constructivism and ANT commit a certain reductionism by obscuring the fact that humans can act in unpredictable ways: "Human actants have a richer behavioural repertoire by which they can respond to prescriptions, and humans may have various intentions, beliefs and motivations that may be relevant in the analysis. In a hybrid vocabulary, these differences between humans and nonhumans are obscured in the interest of symmetrical treatment."[25]

In other words, human agency is polysemic by nature; it can have more than one meaning. Symmetrical models of agency that fail to account for this are deficient and, in presenting a limited scope of human agency, might be reducing the scope of human responsibility as well. While agency might be shared or distributed between humans and technology, the responsibility for the effects of technology always rest squarely with us. We might choose to delegate some of our social agency to digital networks. We might even be compelled to surrender that agency. But we cannot surrender or share with technology the ensuing responsibility for the impact that these actions have on our world.

The point about agency and responsibility is important because, as Paul Dourish points out, the *medium* for acting in the world is increasingly digital, not physical: "[Computer] technology is increasingly the medium within which activity takes place. We are used to the ways in which the physical world mediates our actions, and how it forms a shared environment whose characteristics are thoroughly predictable. . . . Technological systems as a medium for social conduct are very different inasmuch as the inherently disconnected, representational nature of computer systems means that actions can be transformed in unpredictable ways."[26]

Whereas action in the physical world has, by and large, predictable reactions, action in the digital world is mediated in entirely different ways, and agency is assembled in different combinations of human and technological actors—even if humans remain entirely accountable for the consequences of all actions. Since the physical and the digital world are not two separate and discreet dimensions of reality, but are tightly interwoven and interdependent, new models of agency have led to a significant reconfiguration of the self in collectivity.

Methodological approaches like ANT help us map complex connections and dependencies, delineating the political relationship between various actors. They can also help us describe the nuanced ways in which we participate not just in one digital network but in multiple ones. The danger, however, is that these methods are so useful in constructing an approach to studying social realities, that it has become difficult to talk about the network as a singular episteme. For fear of engaging in a form of essentialism, discourses around the digital network remain tied to ideas of multiplicity and plurality, which while valuable also make it difficult to talk about the network as, itself, an essentializing tool of a particular economic and political structure, with concrete implications for how subjects are governed in the so-called networked society.

Governing the Networked Self

Although the social forms that humans and technology have coproduced often appear as innovations, they have never emerged from a historical vacuum. While the digital network reconceptualizes the place of the individual in society, it also replicates many of the features of previous models. Thus in order to understand how the digital network as template produces subjectivity and agency, we need to understand how theories of the modern state define the social contract between the individual, the collective, and the authorities, and how this compares to the models afforded by the network. This exercise can help us determine what exactly is changing as individuals reallocate their agency from one social domain to another, and what this means for democracy. While a complete account of the theoretical evolution of the concept of the modern state is beyond the scope of this argument, some general observations about its nature in relation to the network might be helpful.

The Digital Network and Democracy: Publics versus Masses

If digital networks are said to be transforming participation in everything, including the governance of the state, perhaps it makes sense to begin a comparison of the nature of the state and the network with a discussion of the perceived influence of digital networks on democracy. In general, there are two positions. According to one side, the digital network is believed to be empowering us with new ways of participating in civil society, strengthening our position as a *public*. According to the other, the network is merely a tool of surveillance and regulation, making us more vulnerable to state control, further transforming us into a *mass*. This summary may be overly simplistic, but it is helpful for illustrating some of the tensions surrounding the application of digital networks as tools of democracy. However, a more nuanced reading of the concepts of *public* and *mass*, and how they might be discussed in relation to digital networks, is required.

Previous chapters have already described the shift during our times from a mass society to a network society:[27] from densely knit urban communities that are isolated from each other but organized under the umbrella of the nation-state to a society comprising diffused individuals operating in small sparsely knit communities not bound by location but interconnected by networks. In some of these cases, the transition is imbued with positive connotations, suggesting that the network society represents an opportunity to reverse the formation of masses and return society to the status of a public (mass formation referring basically to a process in which an elite governing class can control the general population, in large part through the dissemination of messages via the mass media). In the network society—the argument goes—digital networks allow individuals to engage in the production of messages, adding their voice to the democratic process instead of being mere consumers of information. This position seems to echo that of philosophers such as Alexis de Tocqueville, John Dewey, Walter Lippmann, C. W. Mills, and Jürgen Habermas,[28] to name but a few, who believe that democracy requires an informed public to operate, whereas nondemocratic forms of government function on the consensual passivity and ignorance of a mass. Most of these philosophers are engaged in a critique of mass culture and mass communication by placing it in direct opposition to a somewhat romanticized notion of the public. Mills, for instance, describes the disparity between publics and masses in terms of three main differences. First, in a public "as many people express opinions as receive them" while in a mass, "far fewer people express opinions than receive them; for the community of publics becomes an abstract collection of individuals who receive impressions from the mass media."[29] Second, in a public "communications are so organized that there is a chance immediately and effectively to answer back any opinion expressed in public"; in a mass, on the other hand, "the communications that prevail are so organized that it is difficult or impossible for the individual to answer back immediately or with any effect."[30] Based on the first two criteria, those who are optimistic about the democratic potential of digital networks can argue that these can facilitate the formation of publics because individuals have increased opportunities for self-expression and can contribute immediate reactions to public discourse with unprecedented effectiveness.

Of course, one can counter this optimism with the arguments of critics who have seen in the dynamics of mass society not the curtailment of self-expression, but its unabated promotion. While recalling the earlier discussion of communicative capitalism, we should remember Deleuze's observation about control societies: "Repressive forces don't stop people expressing themselves but rather force them to express themselves. . . . What we're are plagued by these days isn't any blocking of communication, but pointless statements."[31]

This failure to translate words into action by promoting never ending self-expression brings us to the third and final difference between publics and masses according to Mills. In a public, he argues, "opinion formed by such discussion readily finds an outlet in effective action, even against—if necessary—the prevailing system of authority." On the contrary, in a mass, "the realization of opinion in action is controlled by authorities who organize and control the channels of such action."[32] The question then becomes whether a digital network is an *effective* means for transforming information into meaningful action, or whether—as Rivers[33] proposed—it merely encourages the kind of repetitive, meaningless action that obstructs being.

Old and New Models of Collectivity

Clearly, a comparison of the features of publics and masses in the context of the democratic affordances of digital networks is helpful, but it is not enough to capture the complexity of this new form of collectivity. For that, it might be more productive to compare the digital network to its political predecessor, the modern state, and see which of its features for organizing sociality it adopts, rejects, or reinvents.

The digital network is not making the state obsolete, by any means. But it is, to some extent, giving shape to decentralized and ungovernable multitudes (ungovernable, at least, through the traditional mechanisms of state power, which rely on electoral representation and one-to-many communication). Unlike the state, the digital network is experienced as personal, heterogeneous, fluid, and not bound to a territory. But the state and the network as models of organizing sociality do share some characteristics: they can both be experienced as ubiquitous (the state and the network are all around the individual), and they can both be said to be based on totalizing forms of regulation and mediation based on the dynamics of inclusion or exclusion (one is either inside or outside the state or the network).

The differences and tensions between the state and the digital network might seem insignificant in this current era of globalization and digitality in which both states and networks can be experienced through each other (the network through the state, the state through the network). But philosophically, the debate concerning collectivity and plurality goes back to the very origins of Western modernity and its political theories for conceptualizing the social. One point of departure for this analysis could be Thomas Hobbes and his notion of the state as a great Leviathan. According to Hobbes, a form of government in which individuals subordinate their liberties to a sovereign authority—that is, the state—is necessary and legitimate because, left to their own devices, humans are rather brutish. The quest to satisfy our personal needs and wants means that inevitably we will impinge on our neighbor's needs and wants, resulting in a "war of all against all," which makes our time in this world "solitary, poor, nasty, brutish, and short."[34] Thus it is nothing less than the protection of our lives that the state facilitates. In theory, the state guarantees subsistence, abundance, equality, and security: it makes possible the operation of free markets, declares the equality of all men (literally only *men* at the beginning, since as we know equality for women and other groups considered to be less than white men was only gained through struggles later on), and establishes mechanisms for internal *and* external security (the police and the military, respectively). In return, however, citizens have to enter into a *social contract* in which they recognize that their individual liberties are circumvented by the power of the state. It is worth noting that digital networks are seen as extending the same guarantees that the state offers: in an information economy, they ensure the subsistence of many; their ability to distribute goods easily (some would say *too* easily) guarantees abundance; networks do not discriminate on any bases other than access, so equality is supposedly achieved; and there are governing powers that regulate the networks in order to guarantee our security within them.

Thomas Hobbes's ideas were adopted and adapted by other Western thinkers (Locke, Bentham, Mills Sr. and Jr., and Rousseau, for instance) who helped define the state in terms of a more inclusive liberal democracy in which authority was accountable to citizens to a greater degree than originally imagined by Hobbes. But four characteristics remained essential to the definition of the state: the protection of private property, an emphasis on territoriality as a way to actualize the state, the right of the state to maintain a monopoly on violence, and the equality of all citizens in the eyes of the government. It was seen as the primary role of the state to inculcate in its citizens an inviolable respect for private property, which in fact was seen as predating the state. Without respect for private property, the argument went, there could be no civilization. Because the wickedness of man is universal, it is a given fact that others will try to take the property that is not theirs, so a clear territorial boundary has to be marked between those who have pledged allegiance to the state and respect private property and those who do not. In the process of defending those boundaries, the state reserves the right to employ violence to defend its territory from both external and internal threats. Later, as notions of human rights evolved, the protection of the state was extended to cover *all* citizens, regardless of individual differences. This equality meant that, from the perspective of the state, individuals became subsumed under one totalizing category, "the people," which eclipsed all internal differences "through the representation of the whole population by a hegemonic group, race, or class."[35] The resulting formulation of collectivity and governability was expressed in the belief that the people elect their government with a single will, and the government rules on behalf of the people. Differences in the people are subordinated to the fact that they all enter into the same social contract with the state.

In contrast to this Hobbesian model that gave shape to the construct of "the people," Baruch Spinoza proposes the concept of "the multitude," the *many* not as *one* ("the people"), but as *many*. Even in Hobbes's account, the many predate the one; they precede the state—which is what makes them somewhat of a dangerous and unmanageable entity. Hobbes sees them as rejecting unity and flaunting authority. They recognize no sovereign. Hence the need to establish order and respect for private property by replacing the multitude with the more "civilized" and homogenous concept of the state-bound "people." But authors like Michael Hardt, Antonio Negri, and Paolo Virno take Spinoza's ideas about the multitude and revitalize them into a concept that recaptures the importance of difference and diversity in political affairs: "The multitude is composed of innumerable internal differences that can never be reduced to a unity or a single identity—different cultures, races, ethnicities, genders, and sexual orientations; different forms of labor; different ways of living; different views of the world; and different desires. The multitude is a multiplicity of all these singular differences."[36]

According to Virno, this difference is the basis for a more egalitarian politics: "For Spinoza, the *multitudo* indicates a *plurality which persists as such* in the public scene, in collective action, in the handling of communal affairs, without converging into a One, without evaporating within a centripetal form of motion. Multitude is the form of social and political existence for the many, seen as being many: a permanent form, not an episodic or interstitial form. For Spinoza, the *multitudo* is the architrave of civil liberties."[37]

In this view, diversity does not result in fragmentation, mistrust, and chaos but instead opens up possibilities for the kind of collective action that networks (to a greater degree than states, perhaps) would seem to make possible. According to Virno, the multitude is more, not less, universal than the state. This kind of universality, Hardt and Negri argue, stands in opposition to the global dominance that empire carries out through control, exploitation, and constant war. It is a form of globalization that creates networked circuits of cooperation and that makes it possible to retain difference while discovering—or, more exactly, *producing*—the commonality that facilitates communication and action. This social production of commonality stands in opposition to capitalist production, which is why capitalism has responded by trying to appropriate social production through digital networks organized as monopsonies.

Statelessness and Networks

The reworking of the Spinozian concept of the multitude did not materialize out of nowhere but follows on the footsteps of a Marxist critique of the state, which sees class divisions as indicative of a separation between the rulers and the ruled. This separation is not based on a social contract, but on the exploitation of workers by those who own the means of production. The modern twist is that in the age of digital networks and monopsonies, this exploitation is experienced as benign, even beneficial, and is extended to the social and cultural production that happens beyond the workplace. Long before the so-called socialism of social media and user-generated content, Marx understood that capitalism must seek to commodify not just the worker's manual labor but their social labor as well. To quote Virno, "[N]obody is as poor as those who see their own relation to the presence of others, that is to say, their own communicative faculty, their own possession of a language, reduced to wage labor."[38]

However, Marxism does not seek the abolition of the state, but rather its transformation to a classless state of democratic socialism. And this is where many contemporary theorists part ways with Marxism. What is interesting about modern theories of the multitude is the way in which many of them propose a move toward complete statelessness: the realization by the multitude that it does not need a state. Jacob Grygiel explains this phenomenon:

> Many of today's nonstate groups do not aspire to have a state. In fact, they are considerably more capable of achieving their objectives and maintaining their social cohesion without a state apparatus. The state is a burden for them, while statelessness is not only very feasible but also a source of enormous power. Modern technologies allow these groups to organize themselves, seek financing, and plan and implement actions against their targets—almost always other states—without ever establishing a state of

their own. They seek power without the responsibility of governing. The result is the opposite of what we came to know over the past two or three centuries: Instead of groups seeking statehood through a variety of means, they now pursue a range of objectives while actively avoiding statehood. Statelessness is no longer eschewed as a source of weakness but embraced as an asset.[39]

According to this argument, it would seem that digital networks make possible the emergence of multitudes who, in turn, undermine the state. Unlike states that can be targeted, networked multitudes are dispersed. Their cohesion is not necessarily based on shared identity traits like alignment with a particular nationality, culture, religion, ethnicity, ideology, and so on, although in some instances the network actually serves to accentuate or intensify one of these traits (e.g., global networks based on a particular form of religious or political extremism). Before, statelessness translated into powerlessness—a group without representation in the state did not have any means to assert its political will. Now, however, the statelessness of multitudes is seen as a source of power. While minority groups could always be oppressed within the state, a network allows these groups to organize and act in ways that subvert state control. Or so the theory goes.

Although it has transformed and continues to transform political action—especially when it comes to how special interest groups confront the state—the stateless network tends to exhibit three important failures when it comes to challenging the authority of the state. First, the spontaneity of collective action can be a powerful means of expeditiously organizing a critical mass of individuals to challenge the power of the state, but this initial momentum can just as quickly dissipate as nodes find that there is little or no commonality to support long-term unity and continuity. Large networks that emerge from one day to the other to oppose the state can be powerful political players, but their very size and growth rate work against them when it comes to the slow and painstaking work of negotiating and producing commonality. In short, when it comes to a network's impact, it is "easy come, easy go." Second, the dynamics of network growth (specifically "preferential attachment" in which rich nodes with many links get richer as new nodes link to them) means that the selection of messages and ideas that have the potential to reach large audiences may be more decentralized but not much more democratic, open, or horizontal than the mechanisms found within the state apparatus. In essence, networks (not unlike states) encourage the emergence of big players engaged in a race to accumulate the most attention, and nodes or players that are more "fit" have an advantage over others. All nodes are not created equally. Lastly, movements seeking to heighten their impact need to rely on for-profit networks to quickly increase their membership, In the process, for-profit networks largest number of users. In the process, these for-profit networks are not only able to capitalize on the activity of stateless networks but also perfectly positioned to collaborate with the state in monitoring, detecting, and—when necessary—purging threats to the state. Thus the privatization of stateless space means increased opportunities for surveillance and control on behalf of the state.

Producing Inequality through Inclusion and Exclusion

Despite their differences, states and digital networks share an interesting similarity of sorts: both rely on a kind of *contract* to organize social collectivity. In exchange for the promise of subsistence, abundance, equality, and security (which is after all just a promise, and may differ radically from what the network actually delivers), citizens sacrifice certain aspects of their individuality (such as their privacy) and "pledge allegiance" to a sovereign authority: in the case of the state, it was the rule of law; in the case of the network, it is the algorithms of network logic itself. This contract not only defines what it means to be a citizen or node but also spells out the parameters for participation. Networked statelessness merely replaces the state with other forms of authority and control. It becomes just as difficult to unthink the network as it was to leave the state, because just like statelessness yesterday, networklessness today means political and strategic insignificance. What Agamben observed of identity and statelessness is equally applicable to identity and networklessness: "A being radically devoid of any representable identity would be absolutely irrelevant to the state."[40] We have exchanged our representable identity as *the people* for our representable identity as *the nodes*: a being devoid of its nodality is absolutely irrelevant to the network.

If the digital network has fallen short of its potential to actualize authentic multitudes, it is perhaps due to its inability to come to terms with its outsides, much in the same way that states failed to come to terms with theirs. What the state (and the network) becomes is a reflection as much of what happened outside it, as inside it. After all, the theories of modern sovereignty briefly discussed earlier were developed "in large part through Europe's relationship with its outside, and particularly through its colonial project and the resistance of the colonized."[41] In other words, the enterprise of defining the European subject (and therefore the European state) happened concurrently with the defining of the non-European subject. The logic of the state was in no small part the result of a system of colonial racism that defined the European self in dialectical opposition to the non-European other.[42] Likewise, the networked self is defined to no small extent in relation to the unnetworked other, except that this time the other is not in a faraway colony, but everywhere the network is. Thus both (post)colonial states and digital networks share similarities on how they treat the outside. Depending on the circumstances, the outside or other can be an uncharted domain waiting to be assimilated, a standing reserve waiting to be exploited, a security threat waiting to be diffused, or a combination of those things. The difference is that what is included and excluded in the network is not the result of a demarcated territorial boundary or border, but of a permeable limit that is situated beyond the network as well as between the nodes.

This permeable limit is crucial in unmapping the network, in theorizing how participation not only results in inclusion but also simultaneously results in the exclusion of those who cannot or will not participate and therefore generates inequality. Thus the inequality that digital networks generate revolves around *inclusion* (inequality among nodes within the network), and *exclusion* (inequality between nodes and the outsides of networks). The network as a template for organizing the social creates disparity through enforced participation inside, and exploitation outside. Because of preferential attachment, the rich get richer on the inside. But the wealth of the network is also premised on the availability of an outside to exploit and profit from. "Our wealth depends on their poverty."[43]

In the transition from metaphor to template, the network emerges as a logic or episteme that normalizes this inequality. This logic is accepted because we are told that digital networks create more open and equal social structures. In some ways they do, but there are other processes at work. The dual processuality of networks means they can enable both more freedom (more opportunities for participation and expression) and, paradoxically, more repression (new ways of circumscribing, commoditizing, and monitoring or otherwise controlling the parameters for those new opportunities for action). Not only do we see the creation of new public spaces, but we also see these spaces becoming more vulnerable to monitoring and surveillance, data mining, and the commodification of social labor. When what we gain is overshadowed by what we surrender, it becomes imperative to unmap or unthink the whole structure.

II

UNTHINKING THE NETWORK

[A]nd now the excluded . . . whose lands have been robbed of the minerals, for example, which go into the building of railways and telegraph wires and TV sets and jet airliners and guns and bombs and fleets, must attempt, at exorbitant cost, to buy their manufactured resources back—which is not even remotely possible, since they must attempt this purchase with money borrowed from their exploiters. If they attempt to work out their salvation—their autonomy—on terms dictated by those who have excluded them, they are in a delicate and dangerous position, and if they refuse, they are in a desperate one: it is hard to know which case is worse.

JAMES BALDWIN, *NO NAME IN THE STREET*

5 STRATEGIES FOR DISRUPTING NETWORKS

WHEREAS IT TOOK SEVENTY-ONE YEARS for the telephone to reach half of the homes in the United States, it took only ten years for the same portion of households to get access to the Internet.[1] Certainly, the possibilities associated with the Internet—and with digital networks in general—have not run out their course. But regardless of how new or old technologies are, it is always necessary to question their impact in the political and economic planes in which they operate. At some point, it might even be necessary to set about the task of unthinking the way they have shaped us as a way to reverse some of those impacts. In a climate in which digital networks are being lauded for their positive influence, however, this exercise might seem unnecessary and even antiprogressive. And yet in the case of digital networks, authors such as Tiziana Terranova, Geert Lovink, Jodi Dean, Ned Rossiter, Alexander Galloway, Eugene Thacker, Mark Andrejevik, Evgeny Morozov, Joss Hands,[2] and many others have set out to formulate a critical theory of networks, an analysis that exposes the use of digital networks for the purpose of profit, control, and surveillance. But these authors have also attempted to frame possible ways in which the decentralizing potential of digital networks can be leveraged for articulating new forms of resistance and freedom. After all, as Hardt and Negri suggested, "[i]t takes a network to fight a network."[3]

To the extent that the affordances of a technology can be transformed by human agency, there are possibilities for using digital networks in ways that do not generate inequality. For instance, recent events demonstrate that digital networks can play an important role in organizing resistance movements. The tools of one corporation can be used to organize protests against another corporation or sometimes even against that very same corporation. But we should not let some isolated examples obstruct the truth of what the network has become for the majority of its users: not a tool for changing power structures, but a tool for arresting that change through consumerism and entertainment.

Disrupting or unmapping the digital network is not about celebrating what a small group of hackers can achieve with open-source tools, as important as that work might be. It is about dissecting the way in which the digital network is experienced by the rest of us: the millions of web surfers, prosumers (producers-consumers of media), cell phone users, and video gamers. It is about asking whether the imagination and ethics necessary to resist nodocentrism can emerge from the very networks we use. It is about replacing the notion that someone can design a *better* network with the idea that the network model itself needs to be disrupted. If the logic of the network acts as a social determinant that produces inequality, unmapping it is about conceptualizing the *virtual* sites from which to unthink this logic.

The Virtuality of Networks

If, as Latour suggests, digital networks can "make visible what was before only present virtually,"[4] it is because they allow us to recognize new forms of sociality that before only existed as potentialities. But what exactly is virtuality before it is rendered visible, and through what process is it made so? In other words, in what ways is virtuality already *present* even before we can see the digital network? Does it continue to be present afterward? What is it about the digital network as a technology that makes the process of actualization possible, rendering the virtual visible? In order to answer these questions, we need a better grasp of the virtual, something beyond the common understanding of the word usually associated with concepts like virtual reality, which hint at alternate realms of reality distinct from our "real" reality. Earlier, the false distinction between a networked or mediated self and an unnetworked/unmediated self was explored, and this is an opportunity to continue that discussion in terms of the dichotomy between the *virtual* and the *actual*, as they relate to ways of being in the digital network.

As stated before, early attempts to make sense of emerging social formations facilitated by digital networks conceived of virtuality as a space detached from the local and the "real." This alternate or virtual reality was a separate world endowed with a relevancy of its own and with distinct norms and laws. In this virtual space you could pretend to be whoever or whatever you wanted (on the Internet, nobody knew you were a dog, to quote the famous cartoon). Eventually, however, the distinction between the virtual and the real began to disappear as digital networks integrated more and more aspects of our real and virtual lives. Virtuality (as in cyberspace) was no longer merely a site for manufactured alternate identities (although it continued to afford that), but an enhanced social space for the continuation of our offline identities.

Consequently, the concept of virtuality moved away from popular discourse; people stopped talking about their virtual friends and virtual communities and simply referred to them as *friends* and *communities*. From the perspective of network logic, what mattered was simply whether something was a node in the network or not (what I call nodocentrism). Before, virtuality had been positioned as the *unreal*, an alternative to the real or sometimes even the corruptor of the real (Jean Baudrillard,[5] for instance, bemoans the disappearance of the real and its substitution by the simulated, the virtual). But now virtuality ceased to be perceived as a threat to the real. Nonetheless, as I intend to show by relying on the work of the philosopher Gilles Deleuze, the virtual—and its counterpart, the actual—can be employed to *affirm* the real, increasing our understanding and therefore our engagement with it. In other words, the concept of the virtual can be repositioned as a tool for thinking outside the network, and for intensifying it into a different form of reality.

The problem is not that digital networks *virtualize* the social or make it less real. The problem is that by *actualizing* a social reality (making the virtual visible), digital networks rigidify a social structure and foreclose alternatives. The way to solve this problem, I propose, is to continue the process of actualization in a way that *intensifies* social relations and negates the digital network itself. In order to achieve this, however, we need to start from the virtual.

How to define the virtual? Before offering metaphors and analogies to try to explain what the virtual is, it is pertinent to point out that those are bound to be insufficient and inexact because what we are trying to do is define a kind of ontology. The whole point of defining an ontology— a methodical account of being—is to do away with explanations that require further explanations of a higher order. For Deleuze, the virtual and the actual should not be defined by comparison or by association to anything else, because the virtual and the actual are the ontological building blocks of reality. An ontology defines what is given about reality, what is not questioned. "A philosopher's ontology is the set of entities he or she assumes to exist in reality, the types of entities he or she is committed to assert actually exist."[6] So any metaphor or analogy used to explain the nature of the virtual might be illustrative but should not be confused with the virtual itself. In fact, there is nothing that can effectively be equated with the virtual. The virtual is unlike anything but itself. In this sense, DeLanda characterizes Deleuze's ontology as *realist*, one that grants reality "full autonomy from the human mind, disregarding the difference between the observable and the unobservable, and the anthropocentrism this distinction implies."[7]

Unlike other realist philosophers, Deleuze tries to do away with transcendental explanations of reality. In other words, while for Deleuze reality is not the product of the human mind, it is also not the product of invisible forces, beings, essences, or ideals. Deleuze's ontology is one of immanence: there are no explanations that point to other ultimate realities. Claire Colebrook observes, "In contrast to transcendence as an 'ethics of knowledge' where we seek to obey some ultimate truth, Deleuze described his own philosophy as an ethics of amor fati: as love of what is (and not as the search for some truth, justification or foundation beyond, outside or transcendent to what is)."[8]

Despite previous attempts to situate virtuality as the antithesis of reality, the opposite of the virtual is the actual, not the real. In fact, virtuality is very much part of reality. In the Deleuzian ontology, the reality you and I are experiencing right now is the result of a transformation (or to be precise, a multitude of ongoing transformations) in which an undifferentiated and abstract virtuality becomes a differentiated and concrete actuality. *Becoming* is the unfolding of this transition, the creative act through which things emerge from virtuality as differentiated individuals or actualities. Reality, as DeLanda puts it, is "a relatively undifferentiated and continuous topological space undergoing discontinuous transitions and progressively acquiring detail until it condenses into the measurable and divisible metric space which we inhabit."[9] Everything that exists, in other words, is an actualization of the virtual. Metaphorically (keeping in mind the caveat about using metaphors to explain virtuality), one could compare virtuality to the undifferentiated mess of subatomic particles and actuality to the unique compounds and organisms that emerge as those particles unite and acquire particularity. Or one could compare virtuality to the infinite set of numbers and actuality to specific numbers such as 4, 29, or 23,628,732. In each illustration, it would be impossible to perceive, *all at once*, the virtuality of all the universe's subatomic particles or all the numbers (although we know such a totality exists, if only conceptually). But it is possible to grasp the actualized manifestations of those sets (a particular object, a particular number). The virtual, to be more exact, is not so much the opposite as it is the counterpart of the actual: it is the *unseen* part of the actual that suggests an *invisible* whole, a whole that is nonetheless very much real—not imaginary, conceptual, or transcendent. The virtual, Deleuze argues, is "[r]eal without being actual, ideal without being abstract, and symbolic without being fictional."[10]

Even though it might appear as if the virtual is the source of the actual, the relationship between the virtual and the actual is not hierarchical; virtuality does not represent a purer or higher state from which the actual is derived as a by-product. The fact is that the virtual could not exist without the actual and vice versa; each owes its existence to the other. To return to the point about immanence, Deleuze's ontology attempts to do away with explanations of reality in which our world is seen as a derivative of higher forms or essences. The virtual, therefore, is affirmed in its reality by the actual, anchored by it in the here and now, participating with it in the same single reality, confirmed with each repetition of the process whereby the virtual becomes actualized in a unique and creative way (referred to by Deleuze as the "event").

Repetition is an important characteristic of this process, since actualization—the transformation of virtual into actual—is not a discreet, once-and-for-all occurrence in the existence of a thing. It is an incessant cycle, which is why we can say that objects are continuously and simultaneously virtual and actual: at all times they have one foot in each realm of reality. One way to understand this by analogy is to think of the actual as *specific* and the virtual as *universal* and to think of objects as simultaneously specific and universal. Insofar as we perceive objects as actual, as concrete, they are specific; but insofar as all objects partake of the same single reality, they are universal or (partly) virtual. Another important aspect of Deleuze's ontology is that one thing's way of being real is the same as another thing's way of being real, even when we are talking about two completely different things such as a plant and a rock, or a rock and an idea. The balance of virtual and actual is the same in all things, so that we cannot say that some things are more virtual than others, or that some things are more actual than others. According to

Deleuze, being is univocal instead of equivocal—it has only one sense, not two or more. Being can be said of all things in the same way: the plant is, the rock is, the idea is. Differences in being are conceptual (a result of how we interpret them), not existential.

I suggested earlier that virtuality is the *unseen* part of the actual that connects it to an *invisible* whole ("invisible" not in the sense that it is not real, but in the sense that we might not be able to contemplate it in its entirety). Now that we have established that being is univocal, the nature of this whole becomes clearer. Let us go back to our numbers example. We know that the number four is real in the same way that the number twenty-nine is real (even though they are different numbers or different actualities). We also know that both actualities refer to a whole, which is the set of all numbers. This set is infinite, so it is impossible to grasp or contemplate it all at once. But we know that it is nonetheless real, because any actualization we can conceive or perceive (e.g., any number) is real. Since being can be said of everything in the same way, all actualities necessarily refer to the same reality, to the same whole. Virtuality is this whole, this *common denominator* that all actuals share. This is why Deleuze often refers to the virtual as the *whole* or the *one*. But we must immediately avoid falling into the trap here of reifying virtuality as a transcendental source of the real. The virtual whole does not function as a pure essence, which generates derivative actualities (as in Plato's ontology). As we said, virtuality is part of the same reality as the actuals; it exists in parallel to them and cannot *be* without them. Thus virtuality is a multiplicity. This might sound like a paradox given that we just implied that the virtual whole is a unity. But what we are saying is that the virtual one can only be said to exist through the actualization of the many. In other words, the virtual one should never be accorded a transcendental existence above or apart from the actualized many. Virtuality does not exist at a higher plane of being from which it reigns over actuality as an abstract unity; it is very much part of the world that we perceive through actualization. The virtual and the actual are equally participating partners in the same single reality.

Repetition was mentioned earlier as a key process in this ontology of immanence. Virtuality is an undifferentiated multiplicity, and objects only acquire differentiating attributes (or singularity) when they become actual. But actualization must *repeat* itself over and over again because the actual is situated in space and is subject to time. Repetition generates difference, so for something to become actualized is for it to be constantly repeated and constantly changing: "For the nature of the virtual is such that, for it, to be actualized is to be differentiated. Each differentiation is a local integration or a local solution which then connects with others in the overall solution or the global integration."[11] Therefore, each local solution is not a static node in a network. Difference is individuation, but an individuation that is momentary and contextual, not permanent (which is why a network cannot be drawn once, but must be "animated," set in motion). Difference should not be defined negatively, "as lack of resemblance" (X is different from Y because it lacks this or that attribute), but positively or productively, "as that which drives a dynamic process"[12] (X1 meets Y1 and results in X2). We shall return to the matter of difference in the next section.

But first, I should clarify why Deleuze's theories on virtuality—condensed hastily in these few paragraphs—might be relevant to our understanding of digital networks. We could think of the process of enabling digital networks as a process of actualization through which social structures become concrete and tangible, and we could think of algorithms as specific actualities that make concrete and tangible specific social processes. For instance, the algorithm of collaborative filtering solidifies or actualizes in a particular way both the technical procedure and the social meaning of what it means to recommend something, and to the extent that this algorithm is propagated en masse by monopsonies, it becomes a dominant construct that precludes alternatives from competing in reality. In this sense, nodocentrism actualizes (makes concrete) social formations that were present only virtually; however, once it does so, it also obscures their virtual origins by foreclosing alternatives. When the network reaches the limits of its own nodes, new possibilities need to be intensified.

From Virtual to Intense

An epistemological exclusivity that eliminates everything but the actuality of the node is a form of reductionism. This form of reductionism rejects the virtuality of possibilities that the outsides of networks can engender. And yet the network can only dictate what is possible within it, not what is possible outside it. Paradoxically, by establishing the limits of what is possible in the inside, the network also delineates a plan for how it is possible to differ from it in the outside—that is, it sets the parameters for what we need to do in order to differentiate ourselves from it. The network (to paraphrase Deleuze) is what separates us from knowing ourselves, "what we have to go through and beyond in order to think what we are."[13] In this sense, the network can be applied in the generation of forms of knowledge that can be used to subvert its own logic: The more the network delimits and specifies nodes in one way (by actualizing certain forms of sociality), the more it makes it possible to unmap those nodes in multiple other ways. And although participation in the network need not reveal the inequalities that monopsonies benefit from, the opposition of what is outside the network—multiplied across sites, moments, and identities—can reveal those inequalities, exposing the tension between nodes and nonnodes. The application of unmapping strategies is what can intensify those tensions, what can drive the logic of the network to its limits. The objective of this process of intensification is, simply put, the production of difference.

The way we interpret the digital network is a continuation of the trend Deleuze found in most of Western philosophy to subordinate being to an essentialist and unchanging identity, a way of making sense of the world that requires a fixed subjectivity. Deleuze believes that much of Western philosophy lacks a way to think difference in and of itself, without subordinating it to identity: one thing's identity makes it different from another, places it in opposition (I am me; you are not me; therefore we are different). This kind of difference "implies the negative, and allows itself to lead to contradiction."[14] Deleuze compares a Hegelian worldview, for instance, in which "the thing differs . . . from all that it is not"[15] to one in which "thanks to the notion of the virtual, the thing differs from itself in the first place, immediately."[16] To bring this back to the example of digital networks, in the first worldview or ontology, the node says to what is beyond its limits: "You are not me"; in the second worldview, there is no such thing as *me* (the node) because *I* am already different from *myself* (I am simultaneously node and outside). In terms of a Deleuzian ontology, nodes would not be said to experience life from their own subjectivity, because instead of life being the result of their subjective or interior experience, *they* are in the interiority of life: "Subjectivity is not ours . . . The actual is always objective, but the virtual is subjective."[17] The subjective self is not simply in the objective world or even outside the world. On the contrary, virtuality is a metasubjectivity from which the self is generated again and again in the form of a fleeting actual objectivity. In other words, actualized nodes are but momentary objectivities, and the network is an approximation of a virtual collective or "holistic" subjectivity.

Encountering the outside of the network therefore entails a process whereby the self becomes other to itself, and is "lost" to virtuality. The virtual does not preclude the existence of the individual, but gives us "a universe where individual beings do exist, but only as the outcome of becomings."[18] This is what Rivers refers to as the *openness of being*: one is not what one is, but what one is not yet.[19] This is why the digital network separates us from knowing ourselves. Our ontological vocation is to constantly reinvent ourselves, although the open-endedness of the process of becoming terrifies us. Networks foreclose this open-endedness.

It is through the affirmation of the immanence of virtuality that we learn to accept that the self "becomes double; both loses itself and creates itself."[20] The ongoing movement from the virtual to the actual generates difference not just between individual things but first and foremost within the self, since the self is forever reflecting the multiplicity of the virtual. To Deleuze, there is no such thing as a completely formed, self-sufficient identity; subjectivities and bodies are merely locations for ongoing actualizations. The outside of the network is the space where the self redeems or regains the virtuality that had been ossified in the node, where it encounters (again and again) the others within itself. The relationship between identity and difference is thus reversed: instead of a continuous and stable identity that produces multiple instances of itself through differentiation, it is differentiation itself that gives form to multiple, ever changing identities. As Colebrook states in summarizing Deleuze's philosophy, "Life is difference: to think differently, to become different, to create differences."[21]

These theories, although somewhat abstract, are necessary for the process of thinking and unthinking digital networks, insofar as the obstruction or the production of difference can be achieved through participatory media. Digital networks, although controlled by fewer and fewer media conglomerates, have become important public spaces, so rejecting them completely is impossible. While fighting the network with other networks might make strategic sense in some cases, ultimately it just creates more of the same (more network logic). Instead, I have been suggesting that we must unthink these networks, create alternatives to nodocentric identity. But how? We know we want something *different*, but we do not know what this looks like just yet. My proposed solution is that in order to unthink these networks, to arrive at different solutions, we need to intensify the digital network to the point that it negates itself. By applying strategies of network unmapping, the actualized nodes encounter the resistance of the outside, and the inequalities of the network are brought into focus. This kind of intensification pushes the logic of the network to its limits, turns it against itself, toward new possibilities for social production, participation and action *different* from what is actualized by the network.

Strategies and Directions of Network Disruption

Of the many options available for engaging in network disruption, let us consider three important ones. First, we should engage in an examination of the paradoxes within network logic. To unthink the digital network is to point out the inherent contradictions in their dual processuality. If the digital network increases participation while producing more inequality, if it affords more freedom while creating more opportunities for control, and if it not only makes possible more proximity but also creates more distance, then it is important that we analyze these paradoxes as a way to expose the faults in the logic of the network.

Another strategy is to engage in network parasitology. To unthink the logic of the digital network is not to refuse to confront the network, pretending it does not exist, but to reimagine one's relationship to it. The relationship of the outside to the inside might then be like that of the parasite to the host, if we consider those arguments[22] about how the parasite inserts itself into the communication network between two nodes—the sender and the receiver—disrupting the flow of information by adding noise (information outside the logic of the system) and forcing the network to adjust to its presence. Network science does have at its disposal a way to talk about noninfluential or secondary nodes: if the centrality of a node can be quantifiably described through the metrics of degree, closeness, and betweenness, these measurements can also tell us how peripheral or secondary a node is within a network. What these metrics cannot tell us, however, is how the network can be disrupted by something *outside* the nodes and yet quite proximal to them. This model of communication can provide the grounds for a new model of identity. Communication *in spite of* noise is replaced by communication *through* noise.

One last strategy to unthink the digital network is to create paralogies. This is a term coined by Jean-François Lyotard[23] from the Greek words *para* (besides, beyond) and logos (*reason*). For Lyotard, reason is not a universal faculty that all humans apply equally across all contexts but a subjective and variable form of knowledge production; thus, for him, paralogy is a movement against the established or conventional way of reasoning. Specifically, paralogy "concerns itself with everything that cannot be resolved within the (capitalist) system. In so doing, this form of resistance works by disrupting the instrumental logic of the modern order, producing, for example, the unknown out of the known, dissensus out of consensus, and with this generating a space for micro-narratives that had previously been silenced."[24] In short, paralogy is a "creative and productive resistance to totalizing metanarratives."[25]

Paradoxes, parasites, and paralogies are thus destructive and creative forms of disrupting the network because they actualize forms of differ-ence inside and outside the network that were previously only virtual. These forms of unmapping, of turning network logic against itself, can be achieved through different actions, such as

- *obstruction* of growth in networks;
- *interference* in the flow of information within networks;
- *disassembly* of networks;
- *simplification* (such as localization or slowing down) of processes, mak-ing large-scale networks obsolete;
- *sabotage*, which results in a loss of resources for monopolies and monopsonies;
- *misinformation*, which reduces the value of social trust in networks;
- *hiding* the presence of things that would otherwise be visible in the net-work (for instance, making web pages invisible to search engines, or anonymizing online activities);
- *revealing* the presence of things that would otherwise be invisible in the network (e.g., unveiling secret documents); and
- *intensification* or turning network logic unto itself until it obliterates the network, as will be discussed in the last part of the book.

Disruption can be manifested across multiple sites and contexts. We could say that it can be located nowhere, elsewhere, and everywhere.[26] As the mere expression of an idea—such as in the pages of this book—disruption is a utopia,[27] a *nowhere* that exists in a theoretical or virtual realm. But when disruption is instantiated in the form of a parasite, for example, we can say that it is a heterotopia,[28] an *elsewhere*, a site where exceptional conditions from those of the surrounding system apply. Fur-thermore, to the extent that every node has an outside, disruption is also an atopia:[29] it is borderless, which means it is *everywhere*. Network disruption can evolve as the pursuit of theory or as the application of strategies based on the observation of different network actualizations. Apart from strategies for disrupting the network, we also need directions in which to apply these strategies. Disruption can upset the established order of the network by providing an escape or retreat from the network, by providing a way of reversing network processes, or by standing still in the face of networked progress. Concerning *retreat*, Virno pointed out that to escape is not to passively avoid conflict: "The breeding ground of disobedience does not lie exclusively in the social conflicts which express *protest*, but, and above all, in those which express *defection* . . . Nothing is less passive than the act fleeing, of exiting."[30] Thus any mechanism that allows the subject to escape the digital network by claiming an identity separate from the network is a way to engage in disruption.

Concerning *reversal*, Langdon Winner said that technology is a "license to forget"[31]; as it makes new actions possible, the old way of doing things is forgotten or the resources necessary to do things without the technology become lost. We can think of these licenses to forget as *foldings*. Any approach that allows us to question the impact of technology on the world and explore ways in which the unintended effect of technology's foldings can be *reversed* is a way to introduce a discussion of morals into our use of technology, as Latour argues: "To maintain the reversibility of foldings: that is the current form that moral concern takes in its encounter with technology. We find it everywhere now in the notion of a recyclable product, of sustainable development, of the traceability of the operations of production, in the ever stronger concern for transparency."[32]

When it comes to technology, reversibility is ethics put into practice. If producing large quantities of paper is depleting our forests, the least we can do is to try to reverse the effects through recycling and consider whether this correction will be sufficient. In the same manner, the social benefits that digital networks bring can be assessed and if necessary (i.e., if there are too many unintended negative outcomes) reversed or unmapped. To this effect, unmapping can also serve as the site for nonaction or *stillness* vis-à-vis the network. Here again Latour reminds us that the critical questioning of technology is about considering the value of slowness, about "preventing too ready an access to ends."[33] This inversion can uncover forces and actors that stand in opposition to network logic.

The goal of these strategies and directions, however, is not to collect and arrange networked or unnetworked subjects into "better" networks but to recognize their diversity, agency, and responsibility. The following inquiries into sites of unmapping attempt to uncover the freedom—and also the tragedy—that exists in the disassembly of networks.

6 PROXIMITY AND CONFLICT

Networked Space

Harish lives in Chennai, India. He works for a U.S. company that has outsourced most of its operations. The company's clients are located in North America, while those who provide them with services, like Harish, are in India. His daily routine is not atypical for someone in similar circumstances. After spending the day training new recruits, the other part of his job begins: "At seven-thirty in the evening, when it's 9 a.m. in New York, he confers with the American banking clients for whom he tailors his training, to insure that he is emphasizing the right skills. And then he turns to a slew of computer-programming challenges that may show management his greater gifts. He often goes home after midnight."[1]

Harish's rhythm of life has to accommodate two environments: Chennai and New York. Both environments can be said to be equally relevant to Harish, and thanks to digital networks, both can be said to feel equally immediate or real to him. However, the coexistence of these two geographic spaces does not come without tensions. Harish worries that what feels near to him is becoming increasingly disembodied, detached from his immediate surroundings: "Already, we are half of the time in New York, just our bodies are left behind . . . I worry that nowadays anything near us seems unimportant, while anything we can't see becomes larger than life."[2] Harish's participation in these intersecting networks shapes his perception of social belonging, making it more conceptual and less determined by geographic location: "Lately, he considered community less a function of roads and roofs and tea shops than of imagination. Even the solid presence of his grandmother could dematerialize at the late-night ring of his cell phone, the urgent summons of American clients. And while his parents rolled their eyes at the constant needs of the world beyond Chennai, Harish saw the calls as tidings of cultural integration."[3]

Detachment from one kind of nearness (the immediate environment) is accompanied by attachment to another kind (the mediated environment), and Harish attempts to integrate the benefits of one while not letting go of what is important for him to retain of the other.

Eliot (a blogger's pen name) lives in Charlottesville, a city in the United States. She is a professional website designer, and one of her leisure activities is to coauthor a blog that used to be called Red Inked, now defunct. According to her personal blog posts, digital networks have also fundamentally redefined Eliot's relationship to the near but in a different way than they have for Harish. Commenting on the fear that the Internet replaces face-to-face with mediated interaction, that it makes distant people and places accessed via the Internet more important than one's immediate surroundings and that it foments antisocial habits, she writes, "I'm not chatting with people in New Delhi; nor am I stuck at the computer, turning pale and cutting my wrist to Emo music. Because of the following lists, all on Yahoo Groups, I've gotten connected to and made friends with people in my local geographical area I would not have otherwise met."[4]

She then lists online discussion groups related to recycling, church activities, and networking with working moms. Instead of severing her connections to the near, digital networks have augmented Eliot's links to what is socially proximal: "So my very busy social life, my identity with the town in which I live, and my sense of community—all have been enhanced if not completely created through the weaving of various strands of the web. I have made more linkages and ties to the people in my immediate vicinity than I ever have done in my whole life."[5]

Of course, Harish and Eliot are—literally and figuratively—thousands of miles apart. It would take a lengthy study to discuss the differences between these two cases and their significance. One could start by considering the history and present position in the world's economy of India and the United States and the particular effect that globalization has had in each location. One could then go on to discuss Eliot's and Harish's social class, cultural background, gender, family structures, professional and personal goals, and so on. All this information would perhaps eventually help us understand what accounts for the distinct impact digital networks are having in each case. We might look at Harish's case and conclude that the spatially near is becoming irrelevant, and digital networks are to blame. But then Eliot's case would prevent us from making such broad accusations. We would realize that we also need to take into account the way our use of these technologies engenders new types of nearness or social relevancy within our immediate surroundings, and how this can contribute to new understandings of the world.

Digital networks have fundamentally transformed our sense of what is near and far. As Silverstone argues, "This dialectic of distance and closeness, of familiarity and strangeness, is the crucial articulation of the late-modern world, and is a dialectic in which the media are crucially implicated."[6] And yet there is anything but certainty about the values that are emerging from this process. We are familiar by now with arguments from both sides: those that praise the new social relevancy that digital networks give to the spatially far (the "death of distance" arguments) and those that critique the loss of social relevancy that digital networks impose on the spatially near (the "devaluation of the local" arguments). Through technological mediation, digital networks make it possible to increase our social inclusivity beyond the normal reach of what our bodies and senses allow. But as the cases of Eliot and Harish suggest, different circumstances can yield qualitatively different results when the kind of mediation that digital networks apply to the spatially far is applied to the spatially near. In some cases, that mediation might engender a decreased relevancy of what is spatially near (a form of social exclusivity), and in others it might engender an increased alignment to it (a form of social inclusivity).

Thus digital networks are reshaping social realities by redefining what counts as proximal or relevant (as Heidegger would say, "The frank abolition of all distances brings no nearness. . . . Everything gets lumped together into uniform distancelessness"[7]). But it would be premature to conclude that people are less socially inclined or have fewer social needs than before. People continue to fulfill their social desires, but they do so through new communicative practices, through new mediations of their social realities. The notion of the near as what is *spatially* proximal is being remodeled into a notion of the near as what is *socially* proximal—what we feel is relevant to us socially, regardless of whether it is spatially near or far. For people on the privileged side of the digital divide, the near is no longer bound by space, but instead is something that is constructed through our participation in digital networks. These networks are not antisocial, but highly social. They do not necessarily attempt to do away with the spatially near (the local) but in fact promise us a renewed relationship with it (in addition to new relationships with the spatially far or the global).

Networked proximity reconfigures distance rather than eliminating it. As Borgmann points out, "Information technology in particular does not so much bring near what is far as it cancels the metric of time and space."[8] Within nodocentric logic, nearness is defined in terms of almost-zero distance within the network and farness in terms of almost-infinite distance outside it. What we have then is a shift from physical proximity to informational availability as the principal measure of social relevance.

What kind of social significance does the local acquire under this redefinition of the near? Surely the body and its surroundings cannot simply vanish, even in the spacelessness of the network. Latour[9] observes that a network remains local at every node. The body is thus the node where the network becomes locally situated; it is what remains after the digital network has been shut off. Even the most immersive virtual reality simulation requires the physicality of the body as interface, a body that remains attached to a material environment from which it derives its sustenance. But although it is not possible to completely disentangle the body from the social forces exerted on it by the local, it is true that "physical closeness does not mean social closeness."[10] In other words, we are capable of denying the local a particular significance, acting as if something nearby is not relevant to us. This is what happens when nearness comes to be defined in terms of informational availability and network inclusion, not physical proximity: the local acquires social significance only to the extent that it can be situated within the network, and only aspects of the local that can be rendered by the algorithm of the network acquire social relevancy.

Fears that a mediated or networked proximity might completely replace the local have been blown out of proportion, and in most cases digital networks have augmented or enhanced (or, at least, become entirely integrated with) the local. In the best case scenarios, what was once far can now be near, and what is near can be reapproached through the digital network; nearness, in other words, encompasses not only new forms of global awareness but also rediscovered local solidarities as well. However, questions regarding who gets access to the network or who gets to control its protocols will force us to continue to ask whether networked proximity can or should be disrupted.

This is important because currently network logic is being used to rationalize a model of progress and development where those elements that are not in the network acquire meaning only by becoming part of the network to the point where *bridging the digital divide* is normalized as a goal across society. The problem is that this form of network assimilation, as a strategy for creating nearness, has commodification as its principal motive, since the function of the digital network in a capitalist society is to collect knowledge from its local sources, transform it into more portable information, and generate value by its exchange beyond the local sites. This was already clearly evident in the knowledge management movement, which relied on technology to extract knowledge from individuals and make it applicable across diverse communities of practice by eliminating the information related to the local context and retaining only what was deemed "functional" (i.e., what could be applied regardless of location).[11]

Since the network derives its meaning from the number and diversity of its nodes, the economy of the network is oriented toward converting more things into nodes (commodification), which can exchange information. As a way to counter this assimilation, unnetworked space can function as a paralogy, a site where the network encounters resistance and friction. The outside thus acts as a barrier to the exchange of information by reminding us that not everything can or should be converted into a node. Thus the answer to the problem of network inefficiencies or digital divides is *not* to add more nodes to the network, or even to lower the cost of access, but to find ways of unmapping it.

We must therefore watch against uncritical impulses to make the network universal and all-inclusive, which is what disciplines such as pervasive and ubiquitous computing are attempting to do in order to "empower" humans. Anne Galloway summarizes the ethos of ubiquitous computing: "[U]biquitous computing was meant to go beyond the machine—render it invisible—and privilege the social and material worlds. In this sense, ubiquitous computing was positioned to bring computers to 'our world' (domesticating them), rather than us having to adapt to the 'computer world' (domesticating us)."[12]

The digital network, however, cannot and should not be rendered invisible. If anything, it should be made more noticeable because it is precisely when we pretend it is not there that we are most prone to surrendering our agency, domesticating ourselves to conform to the networks' epistemological exclusivity. Conditioning ourselves to ignore the unnetworked (by believing that anything in the local can be turned into a node) means that we make the network as invisible as the water in

which the fish lives. It is the ultimate surrender to technological determinism and the commodification of knowledge: the ultimate narrative of
exchange value as the most meaningful measure of things.

The premise behind the discourse of the *digital divide* also needs to
be challenged. Unnetworked space functions as the border in the digital divide, as the limit to how far nearness can be technologized (to ask
whether something should be networked or not *is* to encounter the digital divide). Under the logic of the network, however, the digital divide is
seen merely as something to be overcome. Most of the arguments surrounding the digital divide[13] center on the "problem" of those who have
no access to technology and are therefore not on the network, and what
the role of those who do have access should be in addressing this problem. The digital divide has become a metanarrative in its own right,
establishing that the inevitable goal is more network technology that is
applied to more aspects of our social lives and available to more people.
Only then will the playing field be leveled and true progress achieved, we
are told. I do not mean to suggest that some of the problems of our age
could not be alleviated with more technology or, more accurately perhaps, with a more even distribution of technology. But we should take a
closer look at the meaning invoked by the word *divide*.

The discourse of modernity relies heavily on a divide between modern societies and premodern societies to establish a primacy of the
former over the latter, a primacy defined to a large extent in terms of
technological progress that premodern societies must strive to achieve.
Doreen Massey has argued that this dynamic enacts in *space* what is
assumed to be a lag in *time*: "When we use terms such as 'advanced,'
'backward,' 'developing,' 'modern' in reference to different regions of
the planet what is happening is that spatial differences are being imagined as temporal . . . The implication is that places are not genuinely
different; rather they are just ahead or behind in the same story: their
'difference' consists only in their place in the historical queue."[14]

Thus unnetworked space is construed as a place behind the times
(lagging in terms of progress). Unless the digital network manages to incorporate it into its fold, it shall remain infinitely distant in time and space.

The imperative of network logic demands that the digital divide must
be overcome by converting nonnodes into nodes. The result is what
Lyotard calls a "hegemonic teleculture," always working to bring what is
outside the network into the network, to convert unmediated experience
into mediated experience. To be clear, this is not a hegemonic teleculture

because—as Lyotard argues—only distant things are experienced in the digital network. The network is not antilocal, and it does not "abolish local and singular experience."[15] Rather, the digital network is a hegemonic teleculture because things that take place in proximity are treated *the same way* as things that take place at a distance, ensuring that uniform distancelessness reigns.

While networks can no doubt facilitate new forms of engaging the local, the local approached or mediated through the network is not the same local as before, since only elements in the local that are available through the network are rendered as near. While networks are extremely efficient at establishing links between nodes, they embody a bias against anything that is not a node in the network. This is not the same as saying that the network is antisocial or antilocal; in fact, as was established earlier, the network thrives on connecting nodes, and it does not discriminate on the basis of where those nodes are located (in our proximal or nonproximal environment). But when it comes to mediating our relationship with the local, nodocentrism introduces a form of epistemological exclusivity that discriminates against that which is not part of the network.

Nodocentrism can be applied to space to produce a form of hyperlocality that filters out the unnetworked elements in our environment, making them irrelevant. But it can also be applied in a similar manner to a political conflict. The filtering process whereby those elements that are not in the network acquire relevance only by becoming part of the network can both empower and threaten networked actors engaged in organizing action against authority.

Networked Activism versus Networked Surveillance

As Castells[16] suggests, notions of class struggle are being replaced to some extent by notions of a struggle over self-determination between the individual and the network. In most instances, the most effective response in the struggle against networks has been other networks. Because of the scalability and adaptability that is required in a globalized, fast-paced world, the network model has been recognized as the most viable and effective option for confronting disproportionally powerful opponents (as when, for instance, grassroots networks confront corporate or state networks). Framing political struggle in terms of networks fighting networks—pitting one kind of node against another—makes sense from an "evolutionary" perspective, since networks emerged in response to the power of bigger players:[17] speaking in very broad terms and allowing for some historical generalizations, during the last century we saw how political struggles evolved from power blocks fighting other power blocks (as in the case of the Allies fighting the Axis in World War II, or the USA confronting the USSR during the Cold War), to an intermediate stage where distributed networks organized themselves to fight power blocks. Sovereign states found themselves confronting network actors such as guerilla groups, terrorists, or organized criminals employing new distributed tactics that a traditional army or police (even if stronger in manpower or possessing more advanced technologies) was not prepared to confront. This, in turn, developed into a state of affairs where traditional power blocks had to reorganize themselves into networks in order to engage their opponents effectively, resulting in a new era of *netwars*. This form of warfare is accompanied by increased opportunities to conduct aggression not only through the application of the network as organizing model but also through the use of digital networks as weapons or means of conducting warfare. Examples include actions performed by both state and nonstate actors, ranging from the blocking of access to digital networks (in short term, like the Internet shutdowns during protests in Burma, Iran, the Middle East, and North Africa; or in long term, like Israel's refusal to allocate wireless licenses to Palestinian companies[18]), to other acts of cyberwarfare such as espionage, propaganda, vandalism, and the targeting of public services (such as hacking into power plants).[19]

At the same time, authors such as Hardt and Negri[20] have observed that netwar (networks fighting against monoliths or other networks) has evolved to encompass not only military struggles but also struggles for social justice. To give but one example, consider the Zyprexa Kills[21] campaign in which citizens, journalists, and activists used new collaborative communication technologies such as Wikis to organize themselves into a network that opposed a more powerful network of corporate lawyers, researchers, and executives from pharmaceutical company Eli Lilly attempting to cover up the hazardous side effects of their popular neuroleptic product. In cases such as these, it is hard to argue against using digital networks as an effective (and in some cases, the only viable) tool for activism. But while it is politically necessary at times to oppose networks with networks, the application of this tactic is problematic because it can engender new instances of network logic that make it possible for monopsonies to control the subversive networks.

This is obviously evident in the application of digital networks for surveillance. It should not come as a surprise to most people that we are living in an era in which our online movements are recorded in logs that specify what websites we visit, what we search for, what we buy, who we interact with, and so on. Most of the time, these data are used for commercial and advertising purposes only. But it can also be collected and analyzed for security purposes by governments and authorities. Every online utterance on the Internet thus becomes searchable data that artificial intelligence agents can parse for signs of potential threats. Computational approaches such as the Online Behavioral Analysis and Modeling Methodology (OBAMM)[22] can be employed to track a user's behavior, establish normative patterns, and detect deviations that could signify malicious intent, such as when the account has been compromised or the user has gone rogue. Even when we are not on the web, our bodies can continue to be tracked through digital networks. In the United Kingdom alone, for instance, there are now more than four million surveillance cameras in use.[23] Governments might not have the money to staff enough people to monitor all these cameras, so artificial intelligence systems are being perfected that can identify individuals who look threatening or recognize individuals by their facial features, their manner of walking, and so on (all of which might involve some kind of racial profiling). U.S. defense contractors are helping to develop a video surveillance system in China that can identify and track any individual at any given time within an entire city.[24] In the Netherlands, intelligent systems can listen in on ambient sound in public spaces, such as trains, for signs of angry or alarmed speech.[25] And whereas before the police had to worry about placing a wiretap near potential threats to hear what they were saying, now authorities can turn your cell phone into a live microphone and listen to your conversations without your awareness, even if the cell phone is off.[26] In democratic societies, all this happens with our consent because—we tell ourselves—*we* have done nothing bad and have nothing to hide. But what happens when the criteria for what constitute "bad" behaviors changes in the future and the technology is already in place?

The point is that for every new form of dissent that digital networks make possible, more forms of surveillance also become available. And while digital networks allow activists to quickly recruit thousands of adherents to a cause, it has also become easier to dismiss their collective impact and significance. It is not surprising that governments have become (or were always) immune to online petitions, e-mail letters to representatives, or other forms of online activism. The more responsive governments have merely automated the reply to the automated or form letters their citizens send them, resulting in a perpetual cycle of automated democracy.

But to be fair, as tools for activism, digital networks can be used in ways much more powerful than simply sending an e-mail to government representatives. Digital networks extend the opportunities for dissent that are available to the wired citizen, and the organization and expression of voice and action against authority acquires an unprecedented scale: civic groups can not only recruit online supporters in a short time but also actually place them on the street, focusing their attention on an issue as it develops. Taking advantage of mobile technology, mobs become smart participants in protests and can react in real time to developments on the street. Furthermore, the distributed power of collaborative research transforms regular citizens into journalists as they investigate, correct, expose, publish, and republish information before traditional media knows what is going on. The use of portable multimedia devices that can upload data to the network instantaneously also makes it less possible for authorities to act with impunity while assuming that no one is watching.[27] It would appear as if, in an effort to make a quick buck, monopsonies are providing us with the very same tools that could potentially undermine them.

The Activist as Information Aggregator

In most instances, however, activism is reduced to information sharing. This sharing via digital networks can indeed become an act of civil disobedience, especially if the information negatively impacts the interests of corporations or the state (in some cases, the line between information sharing and copyright infringement or plain criminal action is becoming increasingly contested). But the question is how effective as a form of dissent is the sharing of information, particularly when it sees itself as an ends, not a means. In other words, by reducing activism to information sharing through proprietary network technologies, do we further freedom of speech or simply strengthen the authorities' control over the channels of communication and means of action?

A pertinent case to analyze revolves around the distribution of "the number." In early 2007, somebody cracked and published an encryption key to unlock high definition DVDs, allowing for the unrestricted copying of the discs. The key or code started appearing on various websites. The Motion Picture Association of America (MPAA) and the Advanced Access Content System Licensing Administrator (AACS LA) began issuing Digital Millennium Copyright Act (DMCA) violation notices against these websites, demanding that they remove any mention of the number. Some for-profit social media websites, like social bookmarking service Digg, were served with these notices because their users were publishing the encryption key on their posts or comments. The companies attempted to curtail the publication of the number, but there was a massive reaction from users toward this apparent act of censorship: in typical viral fashion, the more the code was being "suppressed," the more it appeared on social media sites, blogs, T-shirts, videos, and so on.

Companies operating under the Web 2.0 business paradigm (capitalizing on their users' social sharing of information) suddenly realized they were in a vulnerable position: they could not afford to alienate their source of free labor, the members of their network. Digg, for instance, reversed its initial decision to block the publication of the encryption key and in a public relations move said that it would rather "go down fighting than bow down to a bigger company." Given its business model, the company (worth at that time around $200 million) might not have had a choice, as it would be nothing without the free labor of its users. As Andrew Lih writes, "This is quite unprecedented—you basically have a multi-million dollar enterprise intimidated by its mob community into taking a stance that is rather clearly against the law."[28]

There are two interesting observations to draw from this example. First, there is the idea of disseminating information using digital networks, in this particular case social media, as a form of activism or protest. This controversy might have had at its core something rather trivial—a code to hack DVDs. But this did not stop some people from asking whether we could extrapolate some of the lessons and techniques learned to a larger social justice context. For instance, Ethan Zuckerman asked, "What would it take to harness this sort of viral spread to harness the net in spreading human rights information? Can activists learn from the story of the number and find ways to spread information that other-wise is suppressed or ignored in mainstream media?"[29]

This is basically what Julian Assange would be doing a few years later with WikiLeaks. But at the time, Zuckerman's comment seemed to suggest that, since the network infrastructure was already in place, what was missing to turn the dissemination of information into a mobilizing force of dissent in society were both the right kind of information and the right kind of audience.

In the encryption key case, it is clear that the "activists" (described by *Bloomberg Businessweek* as predominantly male, in the IT sector, between their twenties and thirties, and earning around $75,000 a year[30]) were more concerned with issues having to do with technology and freedom of speech than with other social issues. As one blogger remarked, "While most of the blogosphere was atwitter over the tantrums being thrown at Digg, real injustice in Los Angeles was being ignored. After watching this video [of police oppression during the May 1st immigration reform march] I was ashamed to be part of a community (the designers and evangelists of Web 2.0?) which sanctimoniously promotes 'people power' among the spoiled and entitled while disregarding the tightening grip of authority on the poor and disenfranchised."[31]

The question is whether the problem is with the type of activist involved in the number controversy, or with the broader framing of an activist as someone who simply manages information, engaged in what Dreyfus[32] would call a nihilism of endless reflection, which never materializes into action. When activism is defined solely in terms of the exchange of information, we are reducing—not increasing—the options available for shaping the world. The activist goes from being a social actor to a mere intersection of data flows. She possesses more information than ever before (about encryption keys as well as about all sorts of social injustices), but all she can do is replicate and pass on the information.

This brings me to the second observation related to the number. In the end, I think the establishment realized that it would be impractical to try to go after Digg, and that doing so might publicize the controversy even further. This case thus signaled a shift in focus from legally prosecuting social media companies for what their members produce and publish to using the social data generated by these sites to monitor for genuine security threats. No one is naïve enough to conclude that social media corporations are really at the mercy of subversive revolutionaries (despite taunting users who posted the number along with comments such as "Hahaha! I am breaking federal law! Hahaha!"). Instead, I believe the lesson learned from this case is that authorities will ultimately recognize the sanctity of capitalism: they will go after individuals rather than companies, and instead of trying to censor speech in online social networks, they will promote it because this gives them more opportunities to monitor dissent. We are back to Deleuze's observation about control societies: "Repressive forces don't stop people expressing themselves but rather force them to express themselves."[33]

The Activist as Street Protester

We have recently seen how activism via participatory media has taken up more consequential causes than making it easier to copy DVDs. In her *New York Times* article "Revolution, Facebook-Style" (published in 2009, before the uprisings of the Arab Spring), Samantha M. Shapiro helps the public visualize what it means "to have a vibrant civil society on your computer screen and a police state in the street."[34] Specifically, she reports on the use of Facebook in Egypt as a means to organize acts of political dissent.

Since 1981, Egypt was ruled under a state-of-emergency law, which severely limited freedom of speech and movement. In 2009, an estimated eighteen thousand citizens were in prison because of this law.[35] In a country where the expression of dissent has such severe social consequences, it is not surprising that citizens gravitated toward a virtual forum where expression was perceived to be more free, and where they did not have to deal with the rigid hierarchies of political groups. Which is probably why, as the article points out, Facebook attracted a new generation of Internet-savvy young people; it was the first foray into political protest for an otherwise disenfranchised segment of the population. By 2008, one particular Facebook group, the April 6 Youth Movement, had about seventy thousand (mostly young and educated) members. The article recounts the story of how the members of this group used Facebook to organize plans to join a march in solidarity with workers protesting high rates of inflation and unemployment. But since the group's online activities were open and visible to all, members of the Egyptian security forces joined the group and tried to dissuade its civilian members from participating in the protest. In spite of this, organizers decided that the march would go ahead.

During the preparations, a thirty-year-old woman named Esraa Abdel Fattah Ahmed Rashid not only disclosed on Facebook the specifics of where she intended to meet some of her peers before joining the protest but also posted full details about the time, what she intended to wear, and even her cell phone number. With all this information, it was very easy for the security forces to arrest Rashid and others during the events. There were at least three casualties that day during the protest. Afterward, people used the same Facebook group to mount a campaign demanding her release, which fortunately happened quickly. However, to the disappointment of many who felt she did not reflect the conviction of her fellow Facebook activists, she appeared on television in tears to apologize for her involvement in the protest (she later withdrew that apology).

Figuring out what to disclose or not to disclose in their digital networks was (and is) a dangerous lesson for young activists to learn. It is undeniable that the use of social media platforms in 2008 (and even before) contributed to the momentum leading to the Arab Spring in 2011, and to a turning point in the involvement of the public (before the Arab Spring, 67 percent of young people in Egypt were not registered to vote, and 84 percent had never participated in a public demonstration[36]). The question is how much of a contribution the technologies made and what were the after effects of their application. After some initial fascination with the concept, there now appears to be more skepticism than support for the idea that tools like Twitter and Facebook are single-handedly responsible for igniting the Arab Spring movements. As we witness the immense effort and cost in human lives that has gone into uprisings in Algeria, Bahrain, Egypt, Iraq, Jordan, Kuwait, Lebanon, Libya, Mauritania, Morocco, Oman, Saudi Arabia, Sudan, Syria, Tunisia, Western Sahara, and Yemen, we recognize that it takes much more than a social media platform to organize and sustain a grassroots protest movement. Yet the liberal discourse behind the trope of a "Twitter Revolution" (a revolution enabled by digital technologies, which empower oppressed groups) continues to function—especially in Western media and academia—as a utopian discourse that conceals the role of communicative capitalism in undermining democracy. In other words, the meme of the Twitter Revolution may have come and gone, but the ideology that gave rise to it continues to color our ideas about participation and democracy.

Digital networks can aid in the defense of human rights, improve governance, and empower the disenfranchised. But that is not the point.

The point is that while presenting these technologies as the agents of revolution, a critique of the capitalist institutions and superstructures in which these technologies operate—and the manner in which they generate inequality—is obscured. Indeed, the use of social media by activists not only increases opportunities for participation and action but also makes it easier for authorities, with help from corporations, to operate a repressive panopticon. According to a report by the OpenNet Initiative, during the Arab Spring around twenty million users in the Middle East and North Africa experienced the blocking of online political content, which was carried out with the help of Western technologies.[37] To the extent that grassroots movements all over the world continue to rely on corporate technologies to organize and mobilize, we can expect inequality (through participation) to take some of the following forms:

Surveillance and loss of privacy. States can monitor activity within digital networks to identify dissenters and learn of (and obstruct) their plans. This is often accomplished through deep-packet surveillance, filtering, and blocking technologies provided to repressive regimes like Iran, China, Burma, and Egypt by companies like Cisco, Motorola, Boeing, Alcatel-Lucent, McAfee, Netsweeper, and Websense.[38] Recently, a group of Chinese citizens even filed a lawsuit against Cisco, claiming that the technology that allowed the government to set up the Great Firewall of China led to their arrest and torture.[39] That the U.S. government pays lip service to the importance of a "free Internet"[40] around the world, and finances circumvention technologies for activists abroad,[41] all while supporting these companies at home through tax breaks and lax regulation is a serious contradiction.

PSYOPs and propaganda. The U.S. Army is developing artificial intelligence agents that would populate social networking platforms and dispense pro-American propaganda.[42] Dozens of these "sock puppets" could be supervised by a single person, and their profiles and conduct would be indistinguishable from those of a real human being (apparently, because of legal issues, these sock puppets could only be targeted to non-U.S. citizens). A low-budget version of this strategy has already been put into action by the Syrian government, which released an army of Twitter spambots to spread proregime opinions.[43]

Loss of freedom of speech. Companies, unlike states, are not obliged to guarantee any human rights, and their terms of use give them carte blanche to curtail the speech of any user they choose. For instance, Facebook (one assumes under the direction of the British authorities) recently removed pages and accounts of various protesters belonging to the group UK Uncut just before the wedding of Prince William and Kate Middleton.[44] UK Uncut is not a violent terrorist organization but a group that opposes cuts to public services and demands that companies like Vodafone pay their share of taxes.

Suspension of service. For more drastic measures, states (in collaboration with corporations) can simply "switch off" Internet and mobile phone services for whole regions in order to terminate access to the resources activists have been relying on. Vodafone, for instance, complied with the Egyptian government's directive to end cell phone service during the January 25 revolution.[45]

Remote control of devices. Modern cell phones have, for some time, provided the authorities with the ability to use them as wiretapping devices without their owner's knowledge, even when the power is off.[46] They can also be used to track individuals and report their locations. An indication of what else we can expect in the future is a patent, filed by Apple, that allows for authorities to remotely disable a phone's camera.[47] While this is intended to prevent illegal recording at concerts, museums, and so on, we can imagine how effective it would be at protests.

Crowdsourced identification. One reason authorities may want to leave the cameras on is because user-generated media can greatly aid in the identification of subversive agents. At the recent Vancouver riots (which had nothing to do with correcting social injustices and everything to do with sports hooliganism), Facebook, Twitter, and Tumblr users were enlisted in a crowdsourcing attempt to identify miscreants using digital photos and videos posted by onlookers.[48] Similar practices were employed by the Iranian government during the postelection riots of 2009. Websites like http://gerdab.ir were setup to allow regime sympathizers to identify protesters and report them to the authorities.[49]

These kinds of practices confirm Morozov's observation that social media can be used by both sides, not just the side we agree with, and that the sacrifices in privacy may not be worth the gains.[50] This perhaps explains why, at least in the Gulf countries, Facebook usage seems to be diminishing.[51] But as regimes—repressive as well as democratic—learn how to use social media to influence the popularity of certain viewpoints, monitor communication, and detect threats, it seems as if dissent will become possible only in the excluded, nonsurveilled spaces of what is outside the network, away from the participation templates of the monopsony.

Nonetheless, something compels people—including at-risk activists—to continue to participate. As Christian Fuchs's research with a student population demonstrates, there is a sharp discrepancy between people's negative opinions of electronic surveillance and their simultaneous willingness to enter into contracts with corporate providers who do not even make a pretense of guaranteeing the privacy of users. In explaining this form of denial, Fuchs writes, "Although students are very well aware of the surveillance threat, they are willing to take this risk because they consider communicative opportunities as very important. That they expose themselves to this risk is caused by a lack of alternative platforms that have a strongly reduced surveillance risk and operate on a non-profit and non-commercial basis."[52]

From this perspective, governments benefit greatly from the process of media conglomeration that their own deregulation policies promote: the more monopsonies become the only game in town—enticing users with the promise increased freedom of expression and organization—the less options for secure or private communication citizens have and the more they will be exposed to surveillance.

And yet some believe that monopsonies actually provide a degree of protection to small dissenting groups. The reasoning is that if these groups were to create and use their own digital networks (e.g., by running open-source software on their own Internet servers), they could be easily targeted and shut down by the authorities. In contrast (the argument goes), targeting an activist group that uses corporate digital networks is a very visible act that would presumably attract a lot of scrutiny and would require the corporation to do a lot of explaining to the public. Zuckerman calls this the "cute-cat theory of digital activism," because according to him "[a]uthoritarian regimes can't block political Facebook groups without blocking all the 'American Idol' fans and cat lovers as well."[53] Unfortunately, this defense of social networking services is faulty because authorities do not need to shut down the whole network but can target—more easily than ever before—only specific groups and members, as described in the examples discussed earlier.

Digital Networks as Consensus Democracies

Another way the digital network handles dissent is through various mechanisms for processing difference of opinion. The algorithms of the digital network can give form to a consensus democracy that *manages* dissent, instead of engaging it as a complex form of disagreement. The network as a model for organizing sociality engenders a kind of homogenizing consensus that, while embracing and thriving on diversity and innovation, obstructs a true measure of otherness, of alternatives. It processes difference algorithmically instead of allowing for the airing of grievances that the agonism of difference produces.

To illustrate this, we can look at normative models for handling conflict in some collaborative spaces of the digital network, such as the pages of Wikipedia. These discursive spaces are often portrayed as ones that embody and promote diversity of opinion and consensus. Wikipedia pages are "social" texts representing a variety of opinions, all the while achieving consensus through mechanisms such as open editing and collective monitoring and correction. According to its how-to pages, Wikipedia enjoins contributors to adopt a neutral point of view (NPOV), "representing fairly, proportionately, and as far as possible without bias, all significant views that have been published by reliable sources."[54] This is intended to promote an environment where a bias expressed by one user motivates another user to challenge it or try to reframe it by substantiating it with facts. The outcome of these kinds of mechanisms is a text where all difference of opinion can be managed through equal representation. But as Rancière suggests, sometimes it is the opportunity for differences and grievances to be openly expressed and not managed through consensus that creates a democratic environment, one where an authentic (if not equal) encounter with the otherness of the opponent can take place: "Democracy is neither compromise between interests nor the formation of a common will. Its kind of dialogue is that of a divided community."[55] Democracy is the many represented as the many in all their inequality, not the many represented as one consensual whole.[56]

What is detrimental to democracy, therefore, is not the absence of difference but the subordination of difference to consensus. Rancière identifies consensus as a state where the rejection of diversity and authentic otherness is more likely to occur because grievances are repressed instead of aired out in the open: "Grievance is the true measure of otherness, the thing that unites interlocutors while simultaneously keeping them at a distance from each other . . . When the apparatus of grievance disappears, what takes over in its stead is simply the platitude of consensus . . . the pure and simple rejection of the other."[57]

Thus for Rancière a rejection of the other is not the result of a lack of consensus, *but of its very presence.* Consensus makes the meaningful expression of grievances impossible. Without the opportunity to claim that a wrong has been committed, there is no opportunity to negotiate an attempt to correct it. Consensus, then, is the loss of meaningful otherness in the sense that it leads to a total rejection of the other in the political arena, for "otherness can only be political, that is, founded on a wrong at once irreconcilable and addressable."[58] Digital networks have a bias toward creating consensus and eliminating grievances through the management of dissent because this creates information and environments that are more efficient and easier to use. But in doing so, networks also have a bias toward a rejection of authentic otherness, epitomized in the incapacity of nodes to recognize anything but a node. Networks can manage difference only as long as that difference is subordinated to the template of the node, but this leads to a total rejection of the only site—the outside of networks—from which authentic grievances against nodocentrism can be expressed. And thus, in order to secure the network, the outside must be declared a threat.

Networked Security

If participation within the digital network creates inequality, this inequality does not give rise to much protest or violence. Rather, inequality is produced and accepted peacefully and consensually by network participants. One reason, as we have seen, is that participation—even when accompanied by inequality—is experienced as pleasurable. The other reason is that inequality in the digital network is rationalized and justified through the *fear* that the real threat to the node comes from outside the network, not from within. Insecurity lurks beyond the borders or limits of nodes. The threat of this insecurity is so great that it makes participation in and of itself enough of a privilege, and enough of a reward to put up with inequality.

What is it about the outside that motivates such fear and makes inequality so readily acceptable? For one thing, the outside represents the unknowable, that which cannot be rendered in terms of network logic and that which has not been (or cannot be) assimilated by the network. To paraphrase Donald Rumsfeld, the outside represents not just the *known unknown* but the *unknown unknown*—the things the network does not even know it does not know. Another reason for this fear is that the network is threatened by difference. It thrives on diversity and inclusion as long as they can be managed internally, but difference outside the established paradigm leads to a loss of control. The difference embodied by the outside is not simply an affirmation of diversity but an affirmation of grievances, which point to authentic otherness. Finally, the network fears contamination—in particular, contamination by paralogical modes of thinking different from nodocentrism. Minor contamination by the outside is allowed because it lets the system build some defenses against it. Contamination is also allowed because the unnetworked contributes resources that benefit the network, even if this is not openly acknowledged. But apart from these instances, a system of security is put in place because of the threat that, if unchecked, the foreign agent that is the outside can infiltrate, run amok, and subvert the system. The outside represents an idea that is dangerous because it can propagate, contaminate, and challenge the status quo.

But perhaps what nodes fear the most, and what keeps insecurity in such sharp relief for them, is the precariousness of their status within the network. Here, interestingly, we find that the lower the threshold for joining the network, the more pronounced the fear of what remains outside. If total inclusion allows for total exclusion,[59] and "the goal of [network] protocol is totality, to accept everything,"[60] *what could possibly be the nature of that which chooses to remain outside?* The outside must thus be eyed with suspicion, even (or specially) if our identities were formed there. The fact that the outside opts out of totality does not reflect well on the decision to join the inside. If the barriers of entry are relatively low, the reasons why the outside refuses to become a node are nothing short of infuriating (e.g., I personally have been accused, in all seriousness, of being irresponsible for not joining Facebook). Freud's narcissism of minor differences could be at play here: ontologically the node and the nonnode are perhaps not so different, but each thinks they are unique, and since structurally they are worlds apart, their rejection of each other becomes a fundamental divide, not least of all because they call into question each other's existence. Otherness is reduced to a few superficial features. But ascribing such fundamentalist views unto the other similarly pushes one's identity to an extreme. We end up reducing *our* identities to a few superficial and nodocentric features as well.

Such extremism or reductionism impels networks to attempt to secure themselves against radical otherness by strengthening their borders—whatever or wherever they might be. Except that in a network, borders no longer exist only at the edges. Rather, they have been distributed and disseminated. The border is everywhere. The barbarian is not at the gate, but standing next to us. Thus a fear of the outside is transformed into a fear of the inside: generalized insecurity. The most dangerous threats to network security are always internal, not external. They come from citizens, not foreigners. We must recall that the unnetworked is not just outside the network but within it. The terror of this "outside" is the fear that immigrant multitudes will undermine the network from inside. Thus as Sützl writes, "[S]ecurity can only be *secured* by insecurity, that is, its self-affirmation is identical with its self-negation."[61] This means that for security to be validated as a goal, insecurity needs to remain a real and constant threat, which means security is an unattainable objective that necessitates the never-ending production of insecurity. As far as the network is concerned, since there is no longer an outside (because the outside is everywhere), insecurity is an ever-present or ubiquitous threat. The way to "secure" the network is therefore to create a perpetual state of surveillance. To the extent that digital networks have become templates of sociality, they have also modeled the management of security and insecurity. Innovation in methods to exploit network vulnerabilities goes hand in hand with innovation in methods for protecting the network, which is why security experts and hackers do more to secure than to jeopardize each other's line of work (meanwhile, the outside of networks escapes firewalls and refuses authentication, tracking, or encryption because it is masked by the node; it eludes the network by creating something the host cannot rid itself of because it might not even be aware of its presence).

One kind of threat that the nonnodal poses to the network involves things like identity theft, service disruption, or denial of service attacks. But these represent instances of networks fighting networks. Another kind of threat is instead epitomized in the confrontation between the surveillance camera and the veiled face of a Muslim woman, which makes identification impossible and is justified on the grounds of human rights, like freedom of religion (although, strictly speaking, veiling is not a practice ordained by the Qur'an[62]). The confrontation between the high-tech surveillance camera and the low-tech veil exposes the tensions in Western discourses between individual freedom and the need to detect "threats," and between voluntary and compulsive participation (in monitored spaces, in the practice of veiling, etc.). In places like France and Barcelona,[63] this tension has been resolved by attempts to ban the veil in public spaces. The message is perfectly clear: in this age of perennial insecurity, the need to monitor presumed threats trumps individual liberties such as religious freedom.

The question then is whether the outside of the network constitutes an authentic threat to the sovereignty of the network, or whether it exists in a symbiotic or parasitic relationship with it. Is the outside merely the network's standing reserve of otherness, ready to be assimilated at a moment's notice, or does it represent an alternative model of identity that could undermine its essence? Perhaps by looking at the use of the network in modern warfare we can discern some answers to these questions.

War and the Terror of Nodes

As discussed earlier, the character of warfare has generally shifted from centralized blocks fighting more or less similar opponents, to blocks fighting decentralized networks, to—more recently—networks fighting networks. This is a model of *asymmetrical* warfare because it allows smaller, weaker groups (such as terrorist or insurgent groups) to fight stronger opponents. Needless to say, this has not made war any more palatable or "fair." On the contrary, netwar has become increasingly inhumane. When asymmetrical opponents confront each other not only on the battleground but also everywhere the network is, the result is disastrous for civilians. A decentralized form of warfare between unequal opponents is one of the factors that could explain why the casualty rate for civilians has gone up from approximately 10 percent in World War I to about 90 percent in the U.S.–Iraq wars.[64] But the hope is that since network technologies have facilitated the practice of war, unthinking network logic might also represent a strategy to evade or resist netwar. Then again, it might just represent a new stage in decentralized warfare. As a lieutenant general in the U.S. Army observed, "Many of our enemies have learned that the way to fight us is not to use technology."[65]

Participatory War 2.0

A distributed or networked war means that individual computer terminals can be recruited into the war effort. While digital networks are providing many ways for organizing resistance to war, they are also providing plenty of ways—from passive to active—to join the war. Social networking services can be used to conduct sophisticated propaganda campaigns, as in the case of the Facebook app that asked users to donate their status bar to alert others about how many Qassam rockets Hamas was firing from Gaza into Israel during the 2009 conflict[66] (in response, a pro-Palestinian group created a similar Facebook app). Likewise, viral video games can be distributed to help one side in a conflict promote their viewpoint (to give but two examples: the game Muslim Massacre involves an American fighter killing Osama bin Laden, the prophet Muhammad, and Allah, while Raid Gaza! shows the disproportionate effect of the war against the Palestinians).

But if propaganda and video games are not enough, more active forms of involvement are also available. Thanks to software that is easy to download and install, any civilian with a computer and access to the Internet can participate in attacks to the web infrastructure of an enemy country. In an article subtitled "How I Became a Soldier in the Georgia-Russia Cyberwar,"[67] Morozov describes how in less than a day he was able to follow simple instructions and use freely available software on his computer to participate in "distributed denial of service" (DDOS) attacks and other acts of vandalism in the 2008 South Ossetia War. A DDOS attack involves overwhelming a web server (hosting, for instance, a government's website) with individual requests or "hits" in order to make it crash and stop working. Just like the volunteer computing projects that make use of donated computer power to help scan outer space for signs of intelligent life (SETI@home), solve complex mathematical problems (ABC@home), or render sophisticated 3D computer animations (Render-Farm@home), new distributed computing software is allowing people to lend their computers to an effort to bring down enemy networks. There are instances of this kind of software to fit all positions across a political spectrum: a group of Muslim hackers designed a DDOS program called al-Durra (named after Mohammed al-Durra, a Palestinian child shot and killed by Israeli soldiers in 2000), while the Israeli group Help Israel Win developed a voluntary botnet called Patriot.

John Robb suggests[68] that this method of cyberwarfare has two main advantages: (a) there is an immense pool of talent willing to participate from the comfort of their homes; and (b) the military, while benefiting from the efforts, can officially distance itself from the actions of civilian militants. According to Robb, while the United States is lagging behind in adopting such trends, Russia and China are embracing them fully and have developed strong relationships with organized crime that allow them to deploy such attacks while at the same time disavowing their participation.[69] To bring the severity of this form of warfare into context, it should be pointed out that causing a web server to collapse is not as innocuous as the inconvenience of getting a "Server busy. Try again later" error. As blogger Jonah Boswitch points out, the disruption of information infrastructures can result in cascading failures affecting systems that support hospitals, air traffic, financial institutions, and so on.

Telesthetic War and Networks

The Internet had its humble beginnings as a military experiment, but information technologies, networks, and war have a long and common history. One of the primary goals of warfare has been to maximize harm to the enemy while minimizing risk to the self, an effort that requires the capacity to inflict damage at an ever-increasing distance from the enemy. Today, we have perfected the technologies to do this, and in the process redefined what it means to *go to war* (does dropping a bomb from half the world away constitute *going* to war?). The ability to conduct telesthetic warfare (i.e., inflicting damage at a distance) requires a speed in the coordination of resources that digital networks and network logic have been developed to provide.

In the heels of the dot-com boom, and using corporations like Walmart and Cisco as models, Vice Admiral Arthur Cebrowski[70] provided the intellectual argument for the idea of network-centric warfare. According to him, information technology networks would revolutionize warfare by bringing digital networks to the military: GPS devices would be ubiquitous, and every soldier would be linked to the network while commands and reports were wirelessly transmitted across the globe. The "fog of war" (an expression that describes the uncertainty that surrounds the battlefield) would finally be lifted. Although initially met with skepticism, this doctrine was vigorously (some say unquestioningly) embraced after 9/11 by the Bush administration. The relative speed and success with which initial missions in Afghanistan and Iraq were accomplished seemed to corroborate this model: thanks in part to superior networks of communication and information, less troops and resources were needed to accomplish preliminary goals (overthrowing Saddam Hussein, for instance). But as initial occupation devolved into participation in lengthy civil wars, the efficacy of the network-centric model began to be contested. As was demonstrated time and again, netwar was not immune to malfunctions: computer systems tended to crash in the heat and the dust, and sometimes there were not even enough battery packs around to power the network. Furthermore, while one can account for all the nodes in one's network, accurately accounting for the forces of the enemy is harder to accomplish. Because of this, a return to telesthetic warfare seems to have displaced the idea of an on-the-ground, network-centric warfare.

As the cases of Afghanistan and Pakistan currently demonstrate, the latest shift in the United States' approach to networked warfare revolves around the application of robotic technologies, in particular unmanned aerial vehicles or drones. These aircraft cost a fraction of what jet fighters cost and can be operated by shifts of pilots thousands of miles away who do not get tired or sleepy. These weapons also depend on very sophisticated digital networks for their operation and guidance. Although a detailed account of the technology is beyond the scope of this text, I do want to at least briefly establish a connection between this technology, network logic, and the ethical repercussions of targeting at a distance. During a drone mission in Afghanistan, a general—in what seems to be a routine episode—ordered a group of civilian houses to be destroyed after video images from a Predator drone showed armed insurgents coming in and out. In the reasoning of the general, not only "was the compound a legitimate target, but any civilians in the houses had to know that it was being used for war, what with all the armed men moving about."[71] The decision to target children, women, and the elderly because they "had to know" that members of their own family are deemed terrorists is a matter that is apparently more expeditiously decided thousands of miles away by looking through a monitor. From a nodocentric perspective, only nodes deserve to be accounted for.

Networks, Social Computing, and Counterinsurgency

Another trend in netwar is to approach the problem of insurgency as a behavior that can be modeled and predicted with social computing algorithms. The intelligence community is asking, "[H]ow can insurgency information best be researched, defined, modeled and presented for more informed decision making?"[72] Social computing, as reviewed earlier, seems perfectly suited for this task since it is concerned with analyzing a social context using algorithms in order to identify patterns and help predict outcomes (in other words, in order to generate a model of the behavior). To date, various approaches are in development, and one of them is STOP, an acronym for SOMA Terror Organization Portal (SOMA stands for Stochastic Opponent Modeling Agents[73]). This online portal allows analysts to access data about terrorist groups worldwide. By hypothesizing a certain state, the system can help analysts predict the behavior of insurgent groups. STOP is composed of the SOMA Extraction Engine (SEE), the SOMA Adversarial Forecast Engine (SAFE), and the SOMA Analyst NEtwork (SANE). A brief description of each component illustrates how network science and social computing can be used for counterinsurgency efforts.

SEE, the extraction engine, uses real-time sources to derive SOMA rules about a particular group. These rules are basically calculations that a computer can perform regarding the actions of a group. SOMA rules take the form of

<Action>:[L,U] *if* <Env—Condition>,

where <Action> is an act (such as kidnapping, arms trafficking, armed attacks, etc.) that the group can undertake, [L,U] is the probability range that this action will take place, and <Env—Condition> is a conjunction of environmental attributes under which the action is likely to take place. In essence, the rule states that "when the <Env—Condition> is true, there is a probability between L and U that the group took the action stated in the rule."[74] For instance, the following rule was derived for the group Hezbollah:

KIDNAP: [0.51,0.55] *if* solicits-external-support & does not advocate democracy.[75]

The rule states that when Hezbollah both solicited external support and did not promote democratic institutions, the probability that they would engage in kidnapping as a strategy was between 51 percent and 55 percent. Similar rules describing behavioral patterns have been extracted from data entered into SEE for twenty-three insurgent groups, including "8 Kurdish groups spanning Iran, Turkey and Iraq, (including groups like the PKK and KDPI), 8 Lebanese groups (including Hezbollah), several groups in Afghanistan, as well as several other Middle Eastern groups."[76] The data for these rules were derived from the larger Minorities at Risk (MAR) dataset developed by the University of Maryland, which tracks the political behavior of 284 ethnic groups worldwide.[77]

Using the data entered into the extraction engine, SAFE (the forecast engine) acts as an online environment where, through the use of drop-down menus, analysts can select a particular group, choose one of the actions available for that group, and select a set of conditions that apply to the hypothetical scenario. For instance, an analyst could ask, "What is the probability at a given time that 'PKK' (group) will engage in 'theft of commercial property' (action) if it 'does not advocate wealth distribution' and it 'solicits external support' (conditions)?" The system then generates the respective probabilities. The last component of STOP and SANE acts as an online social network where analysts can share and discuss various scenarios generated with the system, along with latest news and corresponding background information about the insurgent group from Wikipedia.

The main concern, of course, is how this information will be used. If probability crosses a certain threshold, will certain preemptive actions with "unavoidable" civilian casualties be justified? Incidentally, these concerns are not just relevant to conflicts in the Middle East or Southeast Asia. The U.S. Army has pointed out that SOMA systems can be used domestically to model the behavior of gangs instead of terrorists groups, if one simply substitutes insurgency actions with gang activities.[78] In the aftermath of 9/11, in which we saw the definition of "terrorist" expand to include certain kinds of environmentalists, academics, and other social activists, and at a time when war will increasingly move into urban areas, this does not paint a reassuring picture for voices of dissent even in democracies.

As all these examples show, the real asymmetry in the coming wars will not be between state armies and insurgents but between networks and civilians—or more precisely, between the use of network models to conduct war by states as well as insurgents on the one hand and the civilians that get trapped in this war of network against network on the other. Under these circumstances, the ability to flee or unthink the network will be crucial, and it will become necessary to extend the efforts to disrupt the network to emerging models of collaboration and liberation.

7   COLLABORATION AND FREEDOM

Enmeshed with a global economy, every bit of "free" information carries its own microslave like a forgotten twin.

MATTEO PASQUINELLI, *ANIMAL SPIRITS*

THE TERMS commons-based peer production, social production, Wikinomics,[1] open content, infoanarchism, or as I will simply refer to it here, peer-to-peer (P2P) sharing, may not describe exactly the same thing, but they collectively outline a new model of production and sharing in which people—organized in nonhierarchical digital networks—contribute to decentralized projects, often without financial compensation. The labor generated by the participation of these peers sometimes contributes to a *common good* that is collectively owned by everybody (Wikipedia is a well-known example). But as we saw in the first part of this book, the effort of these peers is increasingly captured and controlled by monopsonies, so that while contributing to the commons is still beneficial, participation in the network produces an inequality that can eventually outweigh that benefit.

The paradox is that the dual processuality of networks drives us toward this outcome while giving the appearance of more, not less, freedom. Thus while in general we have grown accustomed to copyright holders going to extreme measures to prevent the unlawful use of their materials in the production of derivative works (e.g., using a song as background music in a homemade video), we are now beginning to see more "creative" approaches that suggest more freedom, but that also represent more opportunities for corporations to make money. This was abundantly clear in the case of the famous "JK Wedding Entrance" YouTube video, which used copyrighted materials illegally. In the past, the only option in dealing with work uploaded to a digital network that made use of unauthorized components was to remove it. But in the case of this particular viral video, which shows a wedding procession dancing to the beat of Chris Brown's song *Forever* and which was viewed 3.5 million times in the first forty-eight hours after it was released, the record label came up with other options. Thanks to YouTube's automated content identification tool, which scans a file as it is being uploaded for matches to copyrighted work and notifies the owners of the copyright, Chris Brown's label (Sony Music Entertainment, a subsidiary of Sony Corporation) could opt to block or disable the video, track or monitor its views, or monetize the work by choosing to insert advertisements.[2] Instead of blocking the extremely popular video, they decided to embed an advertisement that allowed viewers to click to purchase the song from iTunes.[3] The result was that the song, which had been released a year earlier, enjoyed a revival in popularity and reached top sales spots on iTunes and Amazon. Thus the labor of someone's wedding party and of those who helped the video go viral translated into real profit for the corporations (Sony, Google, its advertisers, etc.), without them having to do much in return.

Some might argue that these are just necessary adjustments to the cost of doing business and living life in the digital age, and that as long as the public gets something out of the deal (entertainment, the ability to easily distribute and share content, fleeting celebrity status, etc.), there is no reason to see in this exchange any sign of exploitation. But what happens when the automated systems like YouTube's content identification tool fail to recognize that the work is being used in a legitimate way (such as applications within the fair use paradigm)? In such cases, the algorithm's inability to deal with nuance might represent a threat to free speech. Furthermore, it is one thing for corporations to be able to make a quick buck from a home movie gone viral. But it is quite another when our very statements about freedom of speech have to be subsidized by corporations due to the fact that without the monopsony our speech cannot reach an audience.

Consider, for example, the Hitler Finds Out meme, a series of YouTube video parodies. The Hitler meme[4] is based on a clip from the 2004 German film *Der Untergang* (*Downfall*, in English), produced by the company Constantin Films. In this particular four-minute scene, Adolf Hitler—played by Bruno Ganz—is informed that Germany is about to lose the war, and he proceeds to have an angry outburst. Video amateurs have taken this bunker scene in German and added fake English subtitles to provide all sorts of new contexts for Hitler to rant about, from technology ("Hitler Finds Out There Will be No Camera in the iPod Touch," "Hitler Finds Out He Has Been Banned from Xbox Live"), to sports ("Hitler Finds Out Newcastle United Have Been Relegated"), to politics ("Hitler Finds Out about Sarah Palin's Resignation," "Hitler Plans to Heckle Barack Obama"), to the more *meta* or self-referential ("What Does Hitler Think of the Downfall Meme?," "Hitler, as *Downfall* Producer, Orders a DMCA Takedown").[5] In early 2010, it was reported that YouTube began to remove some of these parodies at the bequest of Constantin Films, who did not want to be seen as condoning the irreverent parodies (which, according to some, trivialized World War II and the Holocaust). This, of course, unleashed the ire of many infoanarchists and freedom-of-speech advocates, who saw the removal of the videos as a clear violation of fair use (the exception to copyright that allows people to use materials for purposes such as education, news reporting, criticism, and so on without asking for permission). It was not long before a video parody titled "Hitler Reacts to the Hitler Parodies Being Removed from YouTube" appeared.[6] On the one hand, the wit of the authors and the postmodern irony of the parody have to be appreciated. But on the other, one cannot help but wonder at the paradox of allowing a corporation to profit from a product created as a statement against that same corporation's stance against fair use. In other words, why would the authors opt to have a public domain work hosted by a corporation, when they presumably have the tools and knowledge at their disposal to make the work available through other means? Perhaps the answer to that question lies within the parody itself, which has the führer himself deliver the basic lesson behind participation in monopsonies:

> HITLER (speaking about Constantin Film's decision to take down the YouTube videos): "The movie got international attention because of YouTube users' hard work. And now they pull this shit? People worked hard on those videos, and millions of other people loved them! I even made one about Hitler being upset that someone else had taken his Hitler parody video idea! It was fucking great! Now there's nowhere to put it!"
>
> GENERAL BURGDORF: "My Führer, we can probably reupload it to Vimeo or DailyMotion!"

HITLER: "Nobody uses Vimeo or DailyMotion! YouTube is the de facto standard!"[7]

The fact that one corporation functions as the single de facto buyer in the marketplace means that the challenge for the networked participant will be to retain the benefits of collaboration and social production instead of surrendering those benefits to the monopsony. Against the old slogan of "Workers of the world, *unite!*," networked players or laborers may have to find ways to *disassemble*—to disentangle their work from the digital network. The unthinking of networked production will thus involve figuring out how to subvert the alliances that corporations and states are setting up to capture our social and cultural production, so that even action and thought against authority happens along the channels established by them.

(Un)Networked Peers

In theory, peer-to-peer (P2P) networks embody a model of collaboration that, we are told, spells out the end of the monopoly and heralds a new era of equality and creativity. At its most idealistic, discourse on P2P describes a paradigm where all participants are equal and where they voluntarily and freely cooperate with each other in the production of common goods that can be appropriated by anyone, replacing inflexible top-down hierarchies with open modes of production and communication that value reciprocity and sharing over maximization of profit. While the positive impact of successful P2P projects is evident, here I want to contest the status of P2P as an authentic alternative and question some of the norms or values behind the model. The point of this exercise is to investigate whether P2P networks are different from other models of digital networks or whether they merely replicate the same logic. While P2P networks may indeed democratize access to cultural contents, their ability to normalize monocultures needs to be assessed, along with the question of the kind of resistance to hegemony that might be embodied by the peerless, those outside P2P networks.

The Rise of the Digital Commons

While, technically speaking, P2P is just a particular form of network structure, it has come to represent a revolutionary (some would say anticapitalist) mode of production and social organization. What exactly makes this structure so revolutionary? Most digital networks are set up as a system of *servers* that transmit data to *clients*. Some of the advantages of this model are that the distribution of resources is centralized, the production of goods is organized hierarchically, bandwidth is allocated according to one's means to pay for it, and ideas shared within the network can be considered intellectual property and protected by law. In contrast to this centralized architecture, there are no servers and clients in P2P networks because all nodes can simultaneously play the role of server and client as needed. Because there are no dedicated servers, a P2P network has no center.

Because P2P networks still rely on the Internet's basic infrastructure of servers and clients to operate, P2P can be described as a decentralized network structure *superimposed* over a centralized network structure (I will return to this later). What this decentralized structure achieves is the horizontal or *open* production and dissemination of resources, the redistribution of bandwidth according to one's needs through ad hoc connectivity, and the free exchange of ideas unconstrained by intellectual property laws. One consequence of eliminating the distinction between server and client is that peers can engage each other on equal terms: all peers own their own means of production, can access the network in the same way, they have the same opportunities to cooperate, and they all have the same opportunities to derive value from a good. Reward is measured not by profit, but by the opportunities to increase one's knowledge, exercise one's creativity, and increase one's reputation among peers. The result is a commons-based peer production system in which goods can be allocated with no need for monetary compensation; proponents of P2P recognize that digital goods, unlike material goods, can be effortlessly and infinitely reproduced, and it is therefore useless to try to create an artificial scarcity to regulate their exchange.

According to supporters of P2P, the power of collective intelligence behind this model is significantly redefining society at large. Its influence has expanded beyond the open-source and open-content movements to areas like governance, education, science, and spirituality. These changes—we are told—are nothing short of a revolution in moral vision, a "breakthrough in social evolution, leading to the possibility of a new political, economic, and cultural 'formation' with a new coherent logic."[8] Furthermore, P2P is not just ephemeral theory but an actual social practice that signals a major transformation to come.

At a time when the very success of the capitalist mode of production endangers the biosphere and causes increasing psychic (and physical) damage to the population, the emergence of such an alternative is particularly appealing, and corresponds to the new cultural needs of large numbers of the population. The emergence and growth of P2P is therefore accompanied by a new work ethic (Pekka Himanen's *Hacker Ethic*), by new cultural practices such as peer circles in spiritual research (John Heron's cooperative inquiry), but most of all, by a new political and social movement which is intent on promoting its expansion. This still nascent P2P movement, (which includes the Free Software and Open Source movement, the open access movement, the free culture movement and others) which echoes the means of organization and aims of the alter-globalization movement, is fast becoming the equivalent of the socialist movement in the industrial age. It stands as a permanent alternative to the status quo, and the expression of the growth of a new social force: the knowledge workers.[9]

There are, however, some serious limitations behind the idealistic sentiments expressed in this rhetoric. The P2P network is a heterotopia in the sense in which Michel Foucault uses the term:[10] an *other space* with a dual meaning—at once an alternative and a confirmation of the impossibility of alternatives. This is because the "breakthrough" in social and economic evolution that P2P is said to represent is built on top of the same capitalist structures it intends to supersede. For instance, while peers can redistribute bandwidth among themselves, they must first rent it from an Internet service provider (ISP). The production of common goods still depends to a large extent on goods that only some can afford and whose production usually entails some form of exploitation (the production of electronic circuitry used for P2P is still dependent on the surplus labor of the Congolese miner or the *maquiladora* worker).

In short, the decentralization of resources and the deregulation of property are made possible only through the centralization and regulation that capitalism requires. While there are no dedicated servers in P2P networks, information must still flow through a dedicated server at some point because P2P networks are built for the most part on top of the same Internet we all rent from corporations, not a separate Internet. The only reason this world without money is possible is because it is built on top of a world where money is everything. Thus P2P is at once a success and a failure, both a self-sustaining organism and a parasite that cannot live without its host. Baudrillard's observations about simulacra are somewhat useful here: just like he argues that the enclosed space of a prison functions as a convenient way to conceal the fact that the whole of society is carceral, the "free" space of the digital commons that P2P networks create serves to conceal the fact that the rest of the space is subordinated to the logic of capitalism.

P2P might be a rejection of the commodity form, but this rejection is constructed over the old structures of labor and capital that make the commodity form possible in the first place. In capitalism, exploitation happens when the workers, who do not own their own means of production, are made to produce more than what they need to satisfy their needs, and the capitalist uses this surplus labor to generate wealth. Brilliantly, P2P circumvents the model by calling attention to the fact that a surplus of digital goods can be created effortlessly, removing the need for exploitation, and proceeding to facilitate the distribution of tools that puts the means of production into the hands of more people. However, because this happens over a network and socioeconomic structure where not everyone has the resources and knowledge to participate in the digital commons, P2P's "alternative" consists only in a postponement of exploitation: removing it from the pristine sphere of the digital commons by relegating it to other spheres. P2P is, paradoxically, an alternative to the capitalist economy that cannot exist without the capitalist economy. Remove the economy from underneath it—remove the millions of dollars invested in developing microchips and financing warlords that control the mining of Coltan through slavery and rape—and the alternative will cease to exist.

P2P and the "New Socialism"

The desire to buy into the narrative that P2P networks are functional alternatives to capitalism is an expression of a rather romantic form of *digitalism*. According to Pasquinelli, digitalism is "a basic designation for the widespread belief that internet-based communication can be free from any form of exploitation and will naturally evolve towards a society of equal peers."[11] To the extent that proponents of the digital commons (Free or Libre software, commons-based peer production, open source, etc.) believe that digital reproduction can supplant material production in a way that results in more equality (and is better for the environment), they are adhering to a form of digitalism. In the process, unfortunately, they are obscuring the fact that a horizontal democracy of nodes still relies on the surplus labor of an unequal and exploited Other.

Politically, digitalism believes in a mutual gift society. The internet is supposed to be virtually free from any exploitation, tending naturally towards a democratic equilibrium and natural cooperation. Here, digitalism works as a disembodied politics with no acknowledgement of the offline labour sustaining the online world (a *class divide* that precedes any *digital divide*). Ecologically, digitalism promotes itself as an environmentally friendly and zero-emission machine against the pollution of older Fordist modes of industrial production, and yet it is estimated that an avatar on Second Life consumes more electricity that the average Brazilian.[12]

An example of digitalism is the argument that portrays Web 2.0 companies like Flickr and Twitter as the heralds of a new form of socialism.[13] If nothing else, this glorification of the equality-producing qualities of corporate-controlled social media serves to remind us of Virno's observation that, as a way to assuage the revolutionary flames it tends to fan by creating so much inequality, capitalism "keeps providing its own kind of 'communism' both as a vaccine, preventing further escalation, and an incentive to go beyond its own limitations."[14] P2P is part of this process, functioning as an internal communism that makes capitalism seem less savage, as well as a laboratory for the protocapitalist modes of production of tomorrow.

Not for nothing did Virno call post-Fordism the "communism of capital."[15] Post-Fordism is not about the production of material goods in the assembly line, but about the creative production of knowledge and culture through social relations outside the factory. It is the privatization of the public domain. This new form of exploitation, according to Hardt and Negri, translates into "the expropriation of cooperation and the nullification of the meanings of linguistic production."[16] We see it as much in the commercialization of hip-hop as in the adoption of P2P or open source software models by corporations. Big companies have recognized a business opportunity and are plucking the fruits of P2P collaboration in order to reinsert them into the market as commodities. In the name of *social collaboration* and *gift economies*, the users are put to work for corporations. While there are attempts to protect immaterial labor under new collective forms of ownership or "peer property" (licenses like GNU, Creative Commons, etc.), the fact that these models carry within them the ghosts of exploitation cannot be escaped. New models of sociability emerge, but they become organized under a structure where every aspect of the public is owned, hosted, or powered by private interests. A quick look at the terms of use of any social media company will reveal as much. And yet although in essence it is just an experimental expression of private property, peer production is accepted because it gives the illusion (which might be correct superficially) of being more equitable and inclusive. By furthering a capitalist technologizing of sociality, peers are not exactly engaged in the formation of a pure commons, but promote the privatization of collective labor.

Of course, things are not hopeless and P2P is anything but pointless. There are opportunities for resistance and creation in this process. We can respond, as Virno suggests, by "absorbing the shocks or multiplying the fractures that will occur in unpredictable ways."[17] Despite capitalism's attempts to expropriate them, the new models of collaboration opened up by P2P can be fruitful if they are converted into authentic political platforms that revitalize the public sphere. P2P does not have to be a "publicness without a public sphere."[18] It does not have to pose as socialism while increasing our submission to a capitalist order. But for that we might need to think beyond the model of nodes and peers.

The Decline of Cyberpiracy

If there are limits to how much of an alternative to capitalism P2P can be, peers are still beautiful parasites. The heterotopias they create expose the fissures in the system and are testaments to the fact that other ways of thinking are possible. Today, the image in the mirror P2P projects of a world without inequality might be mostly an illusion, but at least it reminds us there is a mirror in which such projections are possible. Nonetheless, a critical assessment of P2P's achievements must continue. While most P2P projects remain small-scale experiments, one phenomenon was cited, until recently, as an example of how P2P could seriously disrupt and threaten the status quo on a mass scale: the piracy of digital goods. Even as the economic impact of digital piracy has been called into question,[19] the cultural significance of this practice remains uncontested. To some, digital piracy conducted through P2P networks is an unavoidable movement toward the redistribution of wealth, making digital goods affordable to audiences who would otherwise not be able to acquire them. According to Nick Dyer-Witheford and Grieg de Peuter, "[M]ass levels of piracy around the planet indicate a widespread perception that commodified digital culture imposes artificial scarcity on a technology capable of near costless cultural reproduction and circulation."[20]

But the rhetoric behind the image of the digital pirate as a cultural and countercapitalist revolutionary leaves some questions unanswered. While global piracy continues to rise, in some countries it is drastically diminishing or at least not growing. According to the RIAA, since 2004 the percentage of Internet-connected households that have downloaded music from P2P networks has not increased. Similarly, a survey conducted by the Business Software Alliance reports that the percentage of youth in the United States who downloaded music, movies, and software without paying declined from 60 percent in 2004 to 43 percent in 2006 and then to 36 percent in 2007[21] (nevertheless, according to the International Federation of the Phonographic Industry, P2P piracy continues to be a problem in other parts of the world like Spain, Brazil, and France[22]). I am neither praising nor lamenting the decline of this form of exchange, nor am I saying there is enough evidence to claim that piracy will eventually disappear or significantly diminish. I am merely suggesting that the largest experiment in P2P adoption seems to be contracting, as strategies and policies begin to reassert the need to conform to social norms and respect private property.

Additionally, it bears asking: if P2P was about empowering individuals to participate in the creation and free exchange of culture, whose culture are most pirates reproducing and circulating with their P2P file sharing clients? Notwithstanding the litany of countercultural practices (hacking, mashing, modding, circuit-bending, speedrunning, etc.) that P2P has facilitated or influenced, the fact remains that for most people, pirating involves the rather uncritical consumption of mass media, the downloading of the latest Hollywood blockbuster or teen idol musical hit. Piracy supplies a tremendous boost to the big artists by popularizing their work, making them even bigger players in the market. The logic of the network reasserts itself: the rich nodes are still getting richer through preferential attachment (new nodes tend to link disproportionately to the nodes with the most links). Digital piracy cannot escape the dynamics that make the network a machine for widening inequalities, not closing them. True, businesses need to adjust to the new dynamics of the industry, but the smart ones will figure out how to capitalize on this "communism." Thus it is incredulous to believe that P2P sharing for the masses will significantly undermine monopolies by creating a *long tail* of diverse cultural alternatives. In an attention economy where traffic equals wealth (even if it is in terms of reputation, not money), the small-time cultural producer can only aspire to become one of the massively shared commodities. Meanwhile, the pirate has only reaffirmed his or her role as a mere consumer in the process. Unlike the piracy of the seventeenth century, this strange form of appropriation or *stealing* only serves to increase the value of the good being *stolen*.[23] The sharing of monocultural goods (and the production of derivatives from these goods) that P2P models facilitate is a form of *ultimate consumerism* in which production becomes the new consumption. It is *ultimate* because (a) social relations outside the market are now commodified through P2P processes and placed inside the market and (b) by remixing monocultural goods and making them available for others to consume, we end up paying for the things we produce. Or as Searls observes in regard to user-generated content, "[T]he demand side supplies itself."[24]

Whereas mass media established a monopoly of communication characterized by the unidirectional flow of information from one to many, digital networks have increasingly come to represent a monopsony of communications where the flow of information is from many to one. Digital networks allow for the sharing of information according to models that seem democratic and egalitarian (models such as the many-to-many P2P), but in terms of the network infrastructure that aggregates and disseminates this information, the model is increasingly that of many users willingly submitting their content to one buyer, who manages it and derives profit from it in unequal proportion (as I argue in chapter 2).

Peers, Outsides, and Disassembly

The P2P paradigm is as nodocentric as any other network model, in that it establishes the irrelevancy of the nonpeer. The "peer" in P2P is an algorithm, a technologizing of a function that solidifies a social interaction according to certain protocols. As cyber peers become capable of recognizing only mirrors of themselves, the labor of nonpeers is externalized from the network and made invisible. But the inequality gap between peers and monopsonies (those who do not own the digital commons, but who still own the physical layer and infrastructure necessary to operate the digital commons) also increases, even as the participation of peers increases. Thus (to come back to the recurring theme of this book) poverty in the network is explained not only by exclusion—as the narratives of the *digital divide* suggest—but also by inclusion under nodocentric terms: it is easier than ever to access and participate in digital networks, but once inside, the logic of the network makes it nearly impossible to escape the dynamic that widens the gap between the wealthy monopsonies and the participating peers.

P2P is also a brilliant failure, but peers do have a supporting role to play in unmapping networked production. P2P allows for the proliferation of parasites in the midst of host systems, and this can serve as the first step in disentanglement. Parasites are useful because they tell us that resistance has conceptualized the first step in unthinking the dominant logic. While parasites or peers may not be able to completely flee the system (they cannot survive without the host), they are able to partially disidentify from the host, to modify the terms of battle. Everywhere the digital network as social template is, commodifying sociality, the peer or parasite can also be, decommodifying the social to a certain extent.

However, this is also where we might encounter the conceptual limits of the peer as a node, and where the resistance of the outside and the peerless becomes important. P2P might be an expression of the will to subvert capitalism, but it is an expression that only exists in one place and always in relation to one entity: the network. It is a commons built on a small corner of the market—the social subordinated to the economy. The peripheries of the network, on the other hand, represent the only sites from which it is possible to unthink the network episteme, helping to conceptualize new models of identity and sociality. Unlike the peer in P2P, the unnetworked aims to be not only inside or outside the host but also where the host no longer *is*. P2P might be a good way of fighting networks with networks, but authentic alternatives will need to contemplate what it means to unthink the network altogether, to find freedom in defection from its logic.

(Un)Networked Freedom

Who would not want the Internet to promote freedom? It is certainly a worthy ideal. Some of the latest proposals made by the Obama administration concerning Internet freedom, however, need to be scrutinized carefully. We need to look beyond the rhetoric of the speeches and examine how actual policies, laws, and joint ventures by the state and the private sector are situating the networked individual in society, and framing the kinds of cultural and social production that can take place within the digital network. Despite the rhetoric, we will find that recent calls from the U.S. State Department in favor of Internet freedom belie a problematic tension between corporate and state interests, on the one hand, and the interests and rights of citizens at home and abroad, on the other.

The story begins with the 2010 fracas between Google and the Chinese government. Most people assume that if you Google something in the United States, you will get the same search results you would get when you Google the same thing in another country. But this is not the case. Countries can and do exert influence on search engine companies to control the results that their citizens can access, and China is a notorious case of this kind of censoring. Doing a search for the word "Tibet" in China, for instance, can yield very different results in that nation than in the West. By early 2010, Google had supposedly gotten tired of the Chinese Communist Party stipulating the kind of search results it could or could not provide to people using its search engine. Google had been doing business in China for some years, and had never expressed any strong concerns over the manner in which the government censored its search results. But things came to a head when it was revealed that the January cyberattacks that compromised the private information of thousands of Google users came from hackers in China, hackers with possible connections to the government. In March, the company decided to stop censoring itself and decided to automatically redirect Chinese users to its search engine in Hong Kong, where everyone could conduct uncensored searches. Soon afterwards, the Chinese government announced it would not be renewing Google's license to operate in the country, which made it seem as if the company would have to leave China later in the year.

After some tension and posturing on both sides, Google and the Chinese government did reach an agreement, and Google's license has been renewed.[25] But at the time, the move to redirect all traffic to the Google Hong Kong site was celebrated in the West as a courageous slap in the face of Internet censorship. Similarly, there were concerns that the possible withdrawal of Google from the Chinese market would make things worse for the average Chinese web surfer.[26] The patronizing assumption was that Google's services were a bastion of freedom inside the Great Firewall of China (one theory for the cause behind the hacker attacks on Google was that the Chinese government resented this freedom and was interested in spying on dissidents' Gmail accounts). This would seem to suggest, to put it plainly, that Google and the rest of the Western IT companies are important tools in the struggle to spread freedom and democracy in China and elsewhere in the world, corroborating a narrative cherished by Western media in which Web 2.0 is bringing democracy to the oppressed world: Facebook liberating Moldova, Twitter aiding a revolution in Iran, and so on. In such cases, social media services are said to have helped mobilize mass protests to contest disputed election results. While not entirely untrue, these claims seem to be exaggerated in the Western media; their purpose seems to be not so much to help contextualize complex social movements as it is to build buzz for the latest social media craze.

To build momentum for this kind of narrative, Google's announcement of its intention to stand up to Chinese censorship was followed, merely a couple of days later, by a speech by Secretary of State Hillary Rodham Clinton on Internet freedom.[27] Because of its importance in facilitating communication and dialogue across various divides, Secretary Clinton (or those responsible for writing her speech) called for an "unfettered worldwide internet," saying, "We stand for a single internet where all of humanity has equal access to knowledge and ideas." In contrast to this vision, she warned that a "new information curtain is descending across much of the world" and critiqued those regimes that are working against freedom: "Some countries have erected electronic barriers that prevent their people from accessing portions of the world's networks. They've expunged words, names, and phrases from search engine results. They have violated the privacy of citizens who engage in non-violent political speech. These actions contravene the Universal Declaration on Human Rights, which tells us that all people have the right 'to seek, receive and impart information and ideas through any media and regardless of frontiers.'"[28]

As examples of what freedom in digital networks should look like, Secretary Clinton remarked that U.S. citizens are free to access any content they want ("Americans can consider information presented by foreign governments") and that people in other countries are free to contact U.S. citizens ("We do not block your attempts to communicate with the people in the United States").[29]

These broad statements seem to completely deny the existence of surveillance, even in the United States, and they also do not reflect the fact that the United States reserves the right to interfere with what other people access in *their own* countries. A case in point would be a recent bill approved by Congress that imposes sanctions on Arab television stations and satellite channels carrying content deemed hostile to the United States.[30] While the bill, HR 2278, intends to censor content produced by Hamas and Hezbollah (which already constitutes an infringement on national sovereignty, according to some), its language is so broad that it actually makes it possible to label a television station a "Specially Designated Global Terrorist" if it airs an interview with someone whose views are considered an "anti-American incitement to violence." As Mark Lynch writes in *Foreign Policy*, "[L]iterally almost every single Arab TV station would be so designated—because no serious Arab TV station could cover the news in the region while ignoring Hamas, Hezbollah, or other figures on the list."[31] The contrast between this bill and the ideals contained in Clinton's Internet freedom speech are evident.

Apart from the issue of who gets to enjoy which freedoms, another problematic set of assumptions about the role of the corporation emerges from the speech. Secretary Clinton expressed the belief that corporations are important champions of Internet freedom in the world: "Increasingly, U.S. companies are making the issue of internet and information freedom a greater consideration in their business decisions."[32] The reality, however, is that some U.S. companies are—as discussed earlier—actively collaborating with both autocratic and democratic governments to use the Internet to monitor and oppress citizens. The public's rights and freedoms always seem to end up taking a backseat to business decisions, both at home and abroad.

In the United States, for instance, the lack of competition (sanctioned by the government) between corporations delivering Internet services impacts the freedoms of the public negatively, as evidenced by recent decisions over net neutrality. Net neutrality is basically the idea that all Internet content should be treated the same and that companies delivering Internet access should not discriminate between different types of content. Thus Internet providers should not be able to charge more or penalize users for downloading certain types of content, for accessing some websites instead of others, or for using particular kinds of software. It would seem that, in this context, recent attempts by the Federal Communication Commission (FCC) to champion net neutrality (hinting of regulations that would ensure transparency and corporate accountability) would be a good thing. But this interest in guaranteeing equal access seems to be destined to succumb to larger corporate interests. In 2008, for instance, the FCC tried to take media conglomerate Comcast to court for intentionally slowing down certain customers' Internet connections because they were using the P2P file-sharing software BitTorrent. However, in April of 2010 a federal appeals court told the FCC it had no right to enforce net neutrality in this manner. To breach this impasse, the FCC proposed that it would legislate Internet transmissions and data separately: while transmissions (how data flows through the wires or airwaves) would be regulated in the same way that wireline phones are regulated, the data itself would be less regulated (just enough to ensure that things like universal service and confidentiality are maintained). This might seem like an optimal arrangement, but what is significant about the outcome is what it represents for the public: a failure to curb monopolies and to promote more competition in the market. In this manner, neither deregulation (allowing monopolies to thrive) nor regulation (applying policies from the wireline era to the Internet) interferes with big corporate interests, and the public is positioned as passive consumers and silent citizens.

Such is the landscape at home; but what about the idea that the corporation is the best candidate for delivering Internet freedom abroad? Unfortunately, there is already a particularly horrendous track record of corporations as champions of American values in foreign lands, as evidenced by the histories of companies like Union Carbide, Dow, Shell, United Fruit, DuPont, Monsanto, and so on. Perhaps comparing Silicon Valley companies to big oil is like comparing Apple to oranges, but let us not forget that some of these IT companies have already been instrumental in helping authoritarian regimes spy on their own citizens or worse (when asked about selling networking hardware they knew the Iranian regime was using to spy on its citizens, a representative from a Siemens–Nokia joint venture replied, "If you sell networks, you also, intrinsically, sell the capability to intercept any communication that runs over them."[33]). In a gesture to public interest, Secretary Clinton did say during her speech that "[t]he private sector has a shared responsibility to help safeguard free expression. And when their business dealings threaten to undermine this freedom, they need to consider what's right, not simply what's a quick profit."[34] Nonetheless, she seems to remain convinced that corporations can help the public express its opinions and organize action abroad, even against authority: "In Iran and Moldova and other countries, online organizing has been a critical tool for advancing democracy and enabling citizens to protest suspicious election results."[35] Unfortunately, online organizing has also jeopardized the privacy and security of activists and opened up new avenues for repression. In this light, there is a troubling side to the partnership between states and corporations in framing the role of the networked activist.

One possible template for this partnership was revealed during the 2009 Alliance of Youth Movements summit in Mexico City. The official goal of the summit was to "explore ways to advance grassroots movements seeking positive social change through 21st century technology and tools," and Mexico was selected because of "its ongoing challenges in addressing violence and crime."[36] Since President Felipe Calderón declared war on drug cartels in 2006, Mexico has experienced an astounding fifty thousand drug-related deaths (it is assumed that most deaths involve criminals, but it is difficult to ascertain how many involve civilians because very few are seriously investigated). Much of this violence is directly attributable to the demand for illegal substances in the United States, and since it is feared that this violence will eventually cross the border, one can understand the motivation for selecting Mexico as a site for a conference on social change. But what is the social media landscape into which Mexican youth are being recruited as activists? A look at the list of sponsors might provide a clue. Although the summit was hosted by Secretary Clinton and the U.S. State Department, it was cosponsored by Facebook, MySpace (at the time owned by Rupert Murdoch), Google, YouTube, Pepsi, MTV, and other corporations. One does not have to be a conspiracy theorist to feel a bit concerned by what seems like the perfect marriage of U.S. foreign policy and for-profit interests, cloaked in a language of liberal democracy, human rights, and social change. In an age when social network analysis is becoming an increasingly important tool for securing the United States, what better way to keep an eye on the volatile youth of the Global South than to have them voluntarily fill out detailed profiles of themselves and of their social networks? From there, the tools of social computing can be applied to try to identify security threats to the network or to engage in the dissemination of propaganda. And if the youth of the Global South can do this while drinking AMP Energy and watching MTV, so much the better, it would seem.

For all its fascination with the "revolutionary" potential of this new form of digital diplomacy, the Obama administration seems to be employing the same failed methods and techniques from past decades. A new generation of young Washington bureaucrats, armed with smart phones and Twitter accounts, have thousands of followers and are able to speak to them in the vernacular of the web (consider two consecutive tweets from Jared Cohen, a member of the State Department's policy planning staff: "Guinea holds first free election since 1958"; "Yes, the season premier [sic] of Entourage is tonight, soooo excited!"[37]). But when time comes to deploy actual strategy, the means and methods seem reminiscent of the one-to-many models of yesteryear, regardless of the new tools. Here is how Alec Ross, senior adviser for innovation to the secretary of state, discusses a social media strategy with Farah Pandith, special representative to Muslim communities for the Department of State:

> You have a body of great material . . . Figure out over the course of whatever it is you've said, those things that can be encapsulated in 140 characters or less. Let's say it's 10 things. We then translate it into Pashto, Dari, Urdu, Arabic, Swahili, etc., etc. The next thing is we identify the "influencer" Muslims on Twitter, on Facebook, on the other major social-media platforms. And we, in a soft way, using the appropriate diplomacy, reach out to them and say: Hey, we want to get across the following messages. They're messages that we think are consistent with your values. This is a voice coming from the United States that we think you wanted to hear. So we get the imam . . . We get these other people to then play the role of tweeting it, and then saying, "Follow this woman," and/or putting it on whatever dominant social-media platform they use.[38]

This top-down approach is seen as an exercise in which American values are translated into a variety of foreign languages and disseminated via the latest media tools. But where are the opportunities to listen to what the audiences in those communities might have to say about those values? No matter how modern the technologies used to deploy it, this "push" model does not seem like a very effective recipe for diplomacy.

The overall assumption behind Clinton's speech is that this model of corporate- and state-sponsored participation in digital networks can not only solve foreign policy problems but also empower world communities socially and economically. Without the slightest sense of irony, she compared the struggle to promote Internet freedom to another infamous revolution: "[T]he internet, mobile phones, and other connection technologies can do for economic growth what the Green Revolution did for agriculture."[39] While this account of history attempts to present the Green Revolution as a technological success, it is difficult to ignore its legacy in terms of the destruction of the environment through pesticides, the impoverishment of our diets, the eradication of native seeds and forms of agriculture, and the increase in world hunger in spite of higher crop yields (due to greed and the disproportionate profits achieved by agro farming corporations at the expense of farmers across the world). One can only hope this is not the kind of success the information revolution has in store for the world.

The examples discussed in this and the previous chapter expose some of the limits of participation in digital networks. In the last section of the book, the idea of intensification as a strategy for unmapping the network, and as a starting point for alternative models of participation, is examined more closely.

III

INTENSIFYING THE NETWORK

Now we have to investigate how the virtual can put pressure on the borders of the possible, and thus touch on the real. The passage from the virtual through the possible to the real is the fundamental act of creation.

MICHAEL HARDT AND **ANTONIO NEGRI**, *EMPIRE*

8 THE LIMITS OF LIBERATION TECHNOLOGIES

DURING THE MOST INTENSE DAYS of the 2011 Egyptian revolution, comedy writer Haisam Abu-Samra wrote about the challenges, and the opportunities, of suddenly experiencing a government-imposed Internet shutdown (in what has become a standard practice during popular revolts, the administration of Hosni Mubarak—in collaboration with Egyptian and Western corporations—suspended access to digital networks in an attempt to diminish the power of activists). While not being able to use mobile phones and web services to communicate with family, friends, and fellow activists contributed to a sense of panic and chaos, Abu-Samra argued that it also brought a clarity of purpose and a reliance on traditional ways of organizing:

> But cutting us out from the rest of the world, from ourselves even, didn't dismantle the revolt. If anything, it removed distraction and gave us a singular mission to accomplish. . . . After suddenly getting thrust into an offline world not only did I learn firsthand how irreversibly entrenched the internet has became in my life and the lives of other Egyptians: I saw how its loss could help us focus our attention on what was happening in reality. The disconnection gave us the chance to prove that we were just as strong, if not stronger, in the face of an authoritarian self-imposed embargo—a decision that itself illustrated the government's fears, not its strengths. . . . Never mind the vacant symbolism of "Twitter revolutions" and Youtube activism: losing the internet at the hand of our own government simply offers us a powerful reminder of why we actually want the internet to begin with, and why we're doing any of this.[1]

Abu-Samra's experiences illustrate what it means to be excluded from digital networks, what it can do to our perception of "reality," and what it might mean in terms of our participation in nondigital networks. However, if we go along with him and quickly dismiss the "vacant symbolism" of Twitter-powered revolutions, and buy into his equally quick embrace of a utopian Internet that empowers citizens and promotes democracy, we might also miss an important opportunity to further clarify and unthink the role of the digital network as a dominant template for organizing sociality. To be sure, the tendency to refer to the Arab Spring movements as "Twitter Revolutions" has thankfully passed. But a liberal discourse of "liberation technologies" (digital information and communication technologies that empower networked communities to change their political realities through mediated participation) continues to influence our ideas about democracy. Unfortunately, this discourse tends to circumvent any discussion of the market structure in which these technologies operate, as if the Internet was not build on a corporate backbone with interests that sometimes run counter to those of citizens.

Even before the so-called 2011 Twitter Revolutions of the Middle East and North Africa, we could point to a series of writings and opinions that suggested that social movements all over the world were being transformed by information and communication technologies (these include, for example, statements about the revolutionary impact of cell phones in the Philippines, YouTube in Iran, Facebook in Moldova, and so on[2]). Stanford University's Program on Liberation Technology captures the idealism behind this movement. On their website, they state that the agenda of the program is to research "how information technology can be used to defend human rights, improve governance, empower the poor, promote economic development, and pursue a variety of other social goods."[3]

While these are noble goals, liberation technology appears to lack an important critical component. Liberation *theology* (which, I am assuming, serves as reference for the concept of liberation technology) sought to lend legitimacy to the struggle of the oppressed by, among other things, questioning the hierarchical structure of the Catholic Church and suggesting that institutions, not just individuals, could be the source of sin and injustice. Unlike liberation theology, however, liberation *technology* does not seem very interested in questioning the roles and structures of the institutions that produce the tools used by popular movements. Instead, liberation technology posits a worldview whereby technologies that emerge in the context of capitalism (precisely at places like Stanford) can be used in the developing or underdeveloped world to bring about social change, presumably in the direction of the kind of democracy that is espoused by the institutions in question.

The discourse of liberation technology tends to present social movements like the Arab Spring as the work of "wired" activists, although this portrayal excludes the work and participation of activists who are not computer literate or simply not social media users. Social change is thus imagined as an outcome of information flows within a network, and activists are portrayed as nodes transmitting dissent to other nodes. In order for liberation to happen, everyone must be connected to the same digital networks. Change and resistance are conceived in nodocentric terms.

Overprivileging a networked view of activism also justifies the export of "subversive" technologies. The discourse of liberation technology accomplishes this by providing two different, although interdependent, versions of the affordances of these technologies: one for the homeland territory and one for abroad. Communicative or information capitalism provides citizens at home no real opportunities for resistance, as the majority of citizens are too occupied compulsively communicating (communicative capitalism is the idea that information and communication technologies materialize ideas of inclusion and participation while subverting resistance to global capitalism[4]). But liberation technology presents a utopian counter narrative of the emancipating and empowering potentiality of technology in places not entirely corrupted by capitalism. This narrative suggests that change, while impossible "here," is realized through liberation technology "over there," in a heterotopian elsewhere (that in the case of the Arab Spring includes the Middle East and parts of Africa). This is a valuable maneuver for liberal sensitivities because it redeems the technologies of communicative capitalism. Activists "over there" are using these tools to talk not just about commercial choices but about things that *really matter*: the overthrow of injustice, the plight of the poor, and so on. Liberation technology thus functions as a form of self-focused empathy in which an Other is imagined who is nothing more than a projection that validates our desires, a user of the same technologies we are using—a hacktivist who applies these tools not for the frivolous ends of consumerism, but for the betterment of the world.

This would seem to imply that the discourse of liberation technology can only serve to arrest social change at home. If that were strictly the case, it would be difficult to account for the Wisconsin protests in early 2011, the emergence of the Occupy movements, or for that matter, any subsequent act of protest in the West that uses technology to mobilize people. The fact that these events continue to germinate and spread seems to demonstrate that it is only a matter of time before social movements influence each other in this age of global media, thus making it possible for liberation technologies to fulfill their true potential wherever the social and economic conditions that fuel social unrest are present, even at home.

What is interesting, however, is that coverage of post–Arab Spring movements in the West has not really revolved around protesters' use of social media, or it has only minimally. Participatory media being used at home for organizing protests is apparently not that newsworthy, since it lacks the sensationalist and media-friendly orientalism of the Twitter Revolution stories. And as the use of participatory media in social movements has become normalized and generalized, there seems to be continued support for the belief that these corporate products have fundamentally shifted the balance of power between producers and consumers and therefore between the owners of the means of production and the audience.

However, I would propose that the discourse of liberation technology conceals, in fact, how production on the new platform continues to exhibit a power imbalance. In theory, the Internet (the über liberation technology in the liberal worldview) brought about the end of communication monopolies with their one-to-many models of dissemination; now, in the age of user-generated content, we have communication that is many-to-many. Access to the tools of production and the channels of distribution has been greatly democratized—the theory goes—and monopolies have been replaced by a free market with perfect competition. Everyone has the opportunity to create content, and everyone has the opportunity to engage that content. While the equation of this continuous communication cycle with civic participation is precisely what the concept of communicative capitalism seeks to critique, we need to also question whether the empowering of more voices has fundamentally changed the monopsonistic market structure of participation.

While the study of resistance movements as networks continues and will continue to be useful, a framework for opposing the nodocentric ordering of these movements into privatized templates for participation is necessary. As activists like Abu-Samra continue to point out to liberation technologists, the struggle must go on after the Internet and other digital networks are shut off. If the fight cannot continue without

Facebook and Twitter, then it is doomed. This means that the struggle is in part *against* those who own and control the privatized networks of participation (and can switch them off at will, or expulse whoever they want). Consequently, we have to turn to sites outside the network for the emergence of corresponding strategies of activism, strategies of intensification that transform online action into offline resistance, and expand the reality of the individual to encompass not just the digital network but the world in both its local and its globalized dimensions.

Alternate Realities

As an educator, I have been exploring one such strategy through the use of alternate reality games (ARG) as platforms for simulation and activism. Although still a work in progress, I have been experimenting with the idea that the digital network can be used in the creation of new forms of knowledge that transcend the limits of network logic, generating ways in which the resistance of the outside of networks can be *intensified* into new models of subjectivity that change what participation means.

ARG are open-ended interactive narratives that are collectively played by participants in real time, using a variety of digital communication technologies such as e-mail, blogs, text messages, digital videos, and so on. Although they have been mostly used by advertisers as tools for viral marketing, they can also be employed to learn about a real-life situation or social problem and imagine different solutions or approaches to it (consider for example the 2007 ARG *World Without Oil*, whose motto was "Play it before you live it"[5]). The objective in this case is not only to raise awareness about a problem in a community but also to collectively propose a number of possible responses to it. This form of networked gaming can thus be framed as a form of participatory action research (PAR), which is concerned with promoting social change through iterative research activities that involve the members of a community. PAR, which has a rich history in Latin America, is a form of collective action through purposeful investigation *by* and *for* the affected community.[6]

In essence, the goal of "serious" (i.e., noncommercial) ARG is to involve communities in analyzing a real-life problem, collectively articulating a multitude of realistic and possible responses to it, and examining the ethical question of *who* has the responsibility to act, and *what* action should look like. Since 2009, I have collaborated with students and faculty at my school to design and deploy an annual campus-wide ARG.

We have addressed themes like budget cuts to the State University of New York system, racism on campus, the local impact of the relationship between Mexico and the United States (in terms of immigration, labor issues, the war on drugs, etc.), Islamophobia, and hydraulic fracturing. While some people would remark (in the case of the ARG that had to do with protesting budget cuts, for instance) that this is merely a replacement of real activism with virtual activism, they would miss the point that, in a depoliticized environment, faculty and students are not engaging in any real activism to begin with.

With this in mind, my students and I put together the following mission statement for our ARG:

> Our mission is to conduct an engaging and interactive Alternate Reality Game to help the SUNY Oswego community address the challenges of possible near-future budget cuts in the context of a state, national and global economic crisis. We seek to involve the community in a constructive dialogue about what we can do, individually and collectively, to prepare our school to meet these challenges. Our focus is on raising awareness, facilitating the generation of solutions, and eliciting action and involvement from members of the college community as well as the city of Oswego and beyond. Additionally, we want to research how new media can be used as a platform for simulation, collective problem solving, and social organizing.[7]

One important aspect of these simulations is how participation is structured. Playing the game is voluntary (or, in some cases, an extra credit opportunity), and students are encouraged to "compete" with one another by completing different levels of participation. These levels range from simply contributing to the online scenarios (participating in the online discussions, and helping to imagine the stories[8]), to higher levels of engagement that transcend the online environment. For example, students can attend on-campus events (lectures, teach-ins, screenings, etc.), actively participate in organizing those events, or organize civic engagement projects related to the theme of the ARG (awareness-raising events, fund raisers, protests, etc.). There is also at least one community forum in which participants get together to discuss the experience and consider the question of what action, if any, they need to take beyond the game.

In this manner, the "virtual" character of these alternate realities is intensified by overcoming the limits of the very networks that give them shape. This is how the unmapping of the digital network takes place; after possibilities have been imagined and explored online, the simulation must be put aside as the community comes together to examine the question, individually and collectively, of what to do next.[9] This completes the passage from virtual to possible to real. From this perspective, ARG can serve to *intensify* social realities, giving shape to something that originates merely as a virtual possibility. Before becoming realities, these possibilities only exist in mediated form; they are language and media constructs that exist merely as bits of information circulating through the network. But these possibilities can be intensified into a concrete reality, a reality that subjects coconstruct through their participation beyond the digital network. If these possibilities were to never transcend the digital network that gave them shape, they would only exist as arrested mediations on the terms that the network dictates. Thus what is interesting to me is not just the medium of the ARG itself (since it is just one strategy of many that could be used to achieve similar ends) but how this medium can be used to generate possibilities that end up negating what is used to create those possibilities in the first place. The goal shifts from the mere actualization of virtualities (making possible new digitized forms of sociality) to figuring out how, in this process of intensification, the digital network itself has become what we have to examine, critique, disassemble, and leave behind—what needs to be negated and disidentified from in order to figure out who and what we are. That is why in future iterations of the ARG, we also want to get students more directly engaged in the production of the online environment, and the questioning of the "liberation technologies" employed to do so.

As we realize that many-to-many communication is becoming impossible without a for-profit many-to-one infrastructure, we must question the narrative that liberation technologies can, by definition, increase democratic participation. Participation managed by monopsony only increases inequality. As networks have become not just metaphors for describing sociality but epistemes that organize and shape social realities, we must examine our investment in networked technologies and the discourses of liberation that accompany them. This way, liberation technology could perhaps be redeemed, if it shifts its focus to using the tools of monopsonies to liberate us from the monopsony itself. But in order to do that, liberation technologists must look beyond the limits of nodes for methods of thinking and acting outside the monopsony.

9 THE OUTSIDE OF NETWORKS AS A METHOD FOR ACTING IN THE WORLD

IMAGINE A NETWORK MAP, with its usual nodes and links. Now shift your attention away from the nodes, to the negative space between them. In network diagrams, the space around a node is rendered in perfect emptiness, stillness, and silence. But this space is far from barren. We can give a name to that which networks leave out, that which fills the interstices between nodes with noise, and that which resists being assimilated by the network: *paranode*. In neuroscience, the paranodal defines a specific type of cellular structure that, while not part of the neural network, plays an important role in excitatory signal transduction. Here, I use the term to refer to the space that lies beyond the topological and conceptual limits of the node. This space is not empty but inhabited by multitudes that do not conform to the organizing logic of the network. As far as the network is concerned, the paranodal exists only to be bypassed or collapsed in the act of linking, of reducing the distance between nodes. But whether it is acknowledged or not, this space gives nodes their history and identity. In other words, the paranodal is not passive; its existence shapes nodes and the relationships between them (much like in urban planning, a "bad" neighborhood "forces" city planners to build a highway around or across it, so that cars can bypass it). The instability of paranodal space is what animates the network, and to attempt to render this space invisible is to arrive at less, not more, complete explanations of the network as a social reality.

To the extent that nodocentrism becomes the dominant model for organizing and assembling the social, only the paranodal can suggest alternatives that exist beyond the exclusivity of nodes. Digital networks create new templates for organizing sociality, but it is only by going beyond the logic of the network that difference from established social norms can be claimed. Furthermore, the paranodal is a site for correcting the nodocentrism that reduces social interaction to self-interested exchange. It is the launching pad for social desires that cannot be contained by the network. These new desires end up causing new shifts and transformations within the network. The paranodal is what forces nodes to react and rearrange themselves according to possibilities that before only existed virtually, causing the network to expand in new directions or even cease to exist. The node, with its static identity and a predefined place and purpose, dissolves into something that can occupy other modes of being and evolving.

The point of conceptualizing the paranodal is not simply to locate and identify what is outside the network in order to bring it within, to assimilate it. Rather, the point is to uncover the politics of inclusion and exclusion encoded in the logic of the network, and to suggest strategies for disidentifying from it. As Rancière suggests,[1] new forms of political subjectification (of shaping consciousness) are always accompanied by disidentification, as certain parts of society reject the whole. The paranodal becomes, to use Rancière's terminology, the part of those who have no part. If digital networks are machines of capitalist subjectification, producing social subjects capable of operating in the privatized pseudopublic space of the network, then it is only in the paranodal where disidentification can take place and alternative subjectivities can emerge.

While the primary directive of the network is linking, paranodality is concerned—to paraphrase Lovink[2]—with whatever the mirror phantom of linking is. A few examples of paranodalities might help to illustrate the concept: a close friend or family member who refuses to participate in the latest social media craze and remains a conspicuous hole in our social network is an example of a paranode; broken web links pointing to pages that no longer exist or cached versions of pages no longer active are paranodal because they represent phantom nodes; signal jammers such as RFID (radio-frequency identification) blockers that prevent network devices from being found are examples of technologies that create paranodality; public spaces without surveillance cameras are paranodal spaces; radio operators without a license (pirate radio) are paranodal because they function without validation from the network; any kind of wilderness where signal reception cannot be established is paranodal; digital viruses and parasites that obstruct the operations of a network are also examples of paranodal technologies; obsolete technology is paranodal because its usage is no longer required to operate the network;

digital noise and glitches are paranodal because they interfere with the flow of data in the network; paranodality is a lost information packet on the Internet; populations in a dataset that are excluded or discriminated against by an algorithm become paranodal; punk or rogue nodes—nodes who belong to a network only in order to destroy it—are paranodal.

Given the multiplicity of networks an individual can belong to at any given time, being paranodal in relation to one network can obviously serve as the basis for belonging to another network. As a starting point, a theory of paranodality can help us account for our participation across these multiple, complex, and open networks. Traditionally, we have thought of the outsides of networks merely in terms of nonmembership, a definite in-or-out status that defines the subject. For instance, Sally Wyatt, Graham Thomas, Steve Woolgar and Tiziana Terranova[3] mapped four types of Internet nonusers: the resisters, the rejecters, the expelled, and the excluded. These categories can be easily transposed to our study of the peripheries of any digital network. The resisters encompass those subjects who have decided voluntarily not to belong to the network; the rejecters used to be nodes in the network but then decided to disidentify from that network voluntarily; expelled nodes also used to be part of a network, but they have been forcefully pushed to the outside; finally, the excluded subjects have always occupied the outside, although not necessarily by their own choice. While these categories are useful for defining what is excluded in terms of a lack of access to the network, they provide too limiting a framework for the construction of manifold networked identities. When it comes to networks, the outside is not just without but within—an outside that is everywhere. The paranodal is a multiversal space that coexists simultaneously with other outsides as well as other insides of networks. It unfolds across various spatiotemporal domains and facets of consciousness. Instead of neatly occupying one of the aforementioned four categories and assuming the corresponding identity, we often find ourselves simultaneously inhabiting a combination of these categories vis-à-vis different networks: one can simultaneously belong to digital technosocial network A, while rejecting network B; find oneself expelled from network C, while continuously resisting belonging to network D; and so on. Furthermore, the peripheries of nodes can involve different kinds of actors (human and nonhuman, material and immaterial) and occupy different topological positions (from the space between nodes, to the borders of networks, to their outsides). Their disassembly can implicate different strategic responses (from passive resistance to active refusal). Each of these possibilities can impact the formation of identity inside and outside the network differently. The point is that across sites, moments, and identities, we simultaneously occupy the place of resisters, rejecters, expelled, and excluded in relation to different digital networks.

A theory of the outside of networks should give us more sophisticated ways to talk not only about nonuse as a mode of disidentification but also about nonparticipation as a mode of resistance. In other words, apart from a more nuanced taxonomy of participation and nonparticipation, the paranodal can help us question the idea of the network itself, in particular with respect to digital networks. Accordingly, the paranodal can provide sites for subverting the idea of the monopsony as the dominant template for our social lives.

Theorizing the outside of networks is about uncovering the paranodal contributions that nodocentrism renders invisible. According to Nick Lee and Paul Stenner, "[W]hatever variable shapes the network may take, the energy required to maintain those shapes is taken, indirectly to be sure, from those who are excluded from the networks."[4] The *wealth of networks*, in other words, is premised on the ability to create systems of exchange that transfer part of the production cost to an external third party: the suppliers of labor, the colonized, the weak, the exploited, and so on. In economics, the term used to describe this deferral is called, aptly enough, an *externality* (e.g., when a company is able to dispose of industrial waste without paying any cleanup costs, this represents an *external* cost to society or the environment). The surplus value that is created by not fairly or fully compensating the paranodal creates the wealth that propels the growth of the network. Even within the network, this wealth disproportionally benefits some parts of the network more than others, which is a way of explaining why in scale-free networks some nodes are more *fit* than others (i.e., they are able to acquire links at a faster rate than others[5]).

It is under these circumstances that the resistance of the outside becomes important. Following David Couzens Hoy,[6] we can say that the resistance that the outside poses to the logic of the inside is an *ethical* resistance because of the kinds of obligations it imposes on nodes. By its mere presence, the outside discloses a site of opposition, making the network aware of the refusal of the unnetworked. Nodes are confronted with a certain obligation to acknowledge the resistance of the outside, even if they opt to actively ignore it or do nothing about it. Nonetheless, this resistance is the only thing that brings the inequalities of the network to the fore. The paranodal can therefore shape the network in very powerful ways, focusing the attention of nodes on the limits of the technosocial systems used to structure their reality. In other words, it is only when nodocentrism is perceived or experienced as an injustice that inequality (between those who participate and those who capitalize on participation) becomes apparent, usually in the form of questions about the politics of network inclusion and exclusion. Through its encounter with the outside, a node can thus run against the limits of its own logic, and be forced to search for horizons beyond its existence and experience as a node in the network.

Standing in the way of such realizations is the fact that the network template has become like the map in the story by Jorge Luis Borges[7] in which a document was drawn with such meticulous detail that it ended up being of the same scale as the territory it sought to depict (in other words, one could overlay the map over the actual space and they would match exactly). Likewise, digital networks do not merely map our current social realities; they organize them and operationalize them so enticingly (promising more friends, more opportunities, and more fun) that the new map replaces the actual territory as the preferred social reality. Thus instead of the map becoming useless—abandoned in the desert like in Borges's allegorical story, populated by the occasional beast and beggar—we increasingly live in the (privatized) network maps created for us.

To talk about disrupting the network under these circumstances may seem like an impossible endeavor. Even if monopsonies are responsible for privatizing and commodifying social relations, it could be argued that they have made sociality more vibrant and interconnected, making it easier (not harder) to express oneself, exercise one's rights, organize against injustice, give voice to minorities, democratize knowledge and cultural production, and so on. By many accounts, the benefits outweigh the costs, making it unrealistic and undesirable to say no to the network. There is much that is valuable in networked participation, and it would be folly to call for its complete rejection. But to engage in a critique of network logic is not to advocate a simplistic form of network rejection. It is to strive to specify the ways in which the network episteme orders our reality. As a philosophical project, disrupting the network is about challenging the determinism of network logic, pointing out the limits of nodocentrism as a form of *othering* that subsumes difference to the contours of the node. As a political project, the point of unmapping the network is to develop the (non)participatory strategies for disrupting the monopsony as a model for organizing the social along profit considerations. Paranodal resistance might take the form of a refusal to do business with certain companies, or a rejection of the premise that we must upload our content to the network with the most users. It might actualize itself as the struggle to get corporations to change their terms of service; or the promotion of open-source, open-content, or peer-to-peer alternatives to monopsonies. It might take messy forms of intensification like the ones Haisam Abu-Samra describes, when Egyptian activists faced an Internet shutdown and were forced to rethink their strategies. Or it might unfold as a form of intensification, which starts within the digital network but moves beyond it, as when some members of the hacker–geek collective Anonymous went from simply "trolling for the lulz" (engaging in various acts of cyber mischief and vandalism just for laughs) to organizing actual on-the-street protests against institutions (the Church of Scientology) and governments (Tunisia, Egypt, Italy, Wisconsin, etc.). According to Gabriella Coleman, the Anonymous "care packet" distributed to participants in the Tunisian operation included language that recognized the limits of cyber activism and encouraged participants to go beyond it: "This is *your* revolution. It will neither be Twittered nor televised or [sic] IRC'ed. You *must* hit the streets or you *will* loose [sic] the fight."[8]

Any kind of project that seeks to give users more control of the data they generate while participating in digital networks should be encouraged: for example, projects that give participants real ownership and portability of their social networking profiles, allowing them to maintain control of privacy settings as they subscribe to various digital networks; or projects that guarantee anonymous searching and browsing of the Internet; and so on. Likewise, the public needs to be better represented when corporations draft the policies that govern their interaction with participants and spell out their rights. The public needs to put pressure on the government to ensure that these agreements are fair, transparent, and binding. Currently, corporations can abuse and exploit users with impunity, and while they are acting within the bounds of legality, a dialogue needs to be started about corporate responsibility in the age of social media. These forms of involvement might not be enough; they merely seek to improve the network rather than unthink it, and they continue to frame participants as somewhat passive recipients of corporate largess—but at least it would be a start.

Perhaps the movement to disrupt digital networks will be akin to what the slow food movement is to fast food: an opportunity to stop and question the meaning of progress. To unthink the digital network would be to constantly decode the relationship between the map and what it represents and the ways in which the map determines or shapes our interaction with the world. Langdon Winner's notion of "epistemological Luddism"[9] might be useful here. Winner argues that we should be able to evaluate technologies based on the following criteria: the degree to which they incorporate participation in their design by the people who will use them, the degree of flexibility and mutability the technologies exhibit (their capacity to be altered and tweaked), the degree of dependency they create, and the degree to which they can be dismantled. But disassembly to Winner is not merely a destructive Luddite reaction to the technology (as justified as that may be, at times). Rather, it is a method, a learning opportunity, a chance to better understand how the technology works, and to better understand how our relationship to it is constituted. This kind of Luddism (what I am calling paranodality as method) might help rogue nodes exploit the entropy that envelops digital networks (an old network is replaced by a newer one; a forced upgrade eliminates a whole category of nodes; users simply stop using a service once the novelty wears off; and so on). In this manner, disassembly would mean the acceleration of the decay of the network, bringing about a reversal of its effects by causing the annihilation of the networked self.[10]

More egalitarian models of social participation might be achieved in the future by challenging the logic of the network. But realistically, today, the paranode might not be able to completely secede from its host and actualize alternatives. As tentative as they may be, strategies like the ones previously suggested can ensure that a critical theory of networks is of practical use to those of us whose social lives are already inexorably intertwined with the services provided by monopsonies. Nevertheless, we should be mindful that none of these proposals and tactics is sufficient or unproblematic. They must be undertaken along with the work of theorizing disidentification from the network, differentiating between what is made possible by the network (the models of participation it affords) and what remains possible only outside of it, and accounting for those parts of the node's own identity that are excluded from the network, preventing it from fully actualizing itself. Thus the scope of what it means to unthink the digital network in the present time should be, beyond the strategies mentioned earlier, to illustrate how the network episteme has molded us, to explain how the network—as cultural metaphor and technological artifact—acts as a social determinant.

Even as we continue to participate in digital networks, we should keep in mind that participation is full of contradictions, and those contradictions define our contemporary existence. In an economy where profit is derived by capitalizing on the participation of users (through advertising, data mining, etc.), and where a handful of buyers acquire and distribute the bulk of user-generated products, great power can be exercised by corporations in setting the conditions under which social exchange can take place. The more participants are willing to accept the conditions defined by the monopsony, the more opportunities there will be for exploitation, and the more the participants will experience an impoverishment as their wealth is reconfigured into immaterial social capital (which is, in any event, managed by the monopsony). An inequality is thus instituted between those who control the network and those who participate in it, an inequality that expresses itself through contradictions: Produce more, own less. Say more, communicate less. Participate more, matter less. Using paranodality as a method means to critique the ways in which the structures of networked participation seemingly make us more versatile actors, while making invisible the manner in which we are being acted on for someone's benefit. In describing the propensity of the public to consume interactive media that creates the illusion of empowerment while solidifying the status quo, Andrejevik observes that "people will not only pay to participate in the spectacle of their own manipulation, but . . . thanks in part to the promise of participation, they will ratify policies that benefit powerful elites and vested interests at their own expense, as if their (inter)active support might somehow make those vested interests their own."[11]

The admission that participation can work against our interests, while seemingly empowering us, should also be a reminder that participation and nonparticipation represent choices laden with values. Increasingly, we will see the question of networked inclusion and exclusion, participation and nonparticipation, framed in ethical terms. For example, students are already being urged by school administrators to forgo participation in some "unethical" digital networks—like the College Anonymous Confession Board[12]—where cyberbullying is prevalent. Similarly, state employees were explicitly told not to participate in the "unethical" WikiLeaks network by reading the released cables, while corporations like Amazon, Bank of America, and Apple[13] also took measures to prevent users from accessing or supporting the "unethical" WikiLeaks through their networks). But apart from considerations of whether it is right or wrong to participate in certain kinds of networks, the resistance of the paranodal must be read in terms of a principled negation of the network. It is only in exclusion (voluntary or involuntary) that alternatives are engendered, and only in exclusion can we find possibilities for disrupting the network, rejecting it, or fleeing from it. Paranodality is nonconformity, and at a time when the logic of the network has found its largest application in privatized systems where the compulsion to participate drives the maximization of profit and endangers the democratization of cultural production, paranodality as method means revitalizing nonconformity as the site of important debates.

Digital networks and the network episteme (the network as a strategy for knowing the world) have already transformed who we are and how we interact with each other—at least for the third of the world's population who have access to the Internet and the 70 percent who have access to mobile phones. It is impossible, perhaps even undesirable, to turn back the clock to a time of pre(digitally)networked societies. Thus the more realistic strategies for unthinking and unmapping networks will rely not on abandoning them in a technophobic reaction; they will rely on the intensification of the network: questioning the terms under which it includes and excludes, engaging in creative acts of disassembly by pushing the limits of its logic, and conceptualizing alternative modes of being through the paranodal. We are just beginning to imagine what disrupting the network might look like.

NOTES

Acknowledgments

1 Deleuze, *Difference and Repetition*, xxi.

Introduction

1 Kiss, "Facebook."

2 Warren, "Quit Facebook Day Falls Flat"; Spring, "Quit Facebook Day Was a Success Even as It Flopped."

3 White, "Facebook Suicide."

4 From the banner at Quitfacebookday.com. Ironically, in 2009 the website Seppukoo (www.seppukoo.com), whose goal was to "assist your virtual identity suicide," received a cease-and-desist letter from Facebook accusing them of malicious appropriation of the personal data of users.

5 Wellman, *Networks in the Global Village*.

6 Deleuze and Guattari's notion of rhizomatic thinking, for instance. See Deleuze and Guattari, *A Thousand Plateaus*.

7 Tryhorn, "Nice Talking to You."

8 Eskelsen, Marcus, and Ferree, *The Digital Economy Fact Book*, 6. The exceptions are Mexico, Turkey, Portugal, Italy, and Ireland.

9 Eskelsen, Marcus, and Ferree, *The Digital Economy Fact Book*, 6.

10 Tryhorn, "Nice Talking to You."

11 Eskelsen, Marcus, and Ferree, *The Digital Economy Fact Book*, 60

12 See for example Scoble and Israel, *Naked Conversations*; Micek and Whitlock, *Twitter Revolution*.

13 See for instance Feld and Wilcox, *Netroots Rising*; Roberts, *How the Internet Is Changing the Practice of Politics in the Middle East*.

14 Weis and Andrews, *The Business of Changing Lives*.

15 Bonk, *The World Is Open*.

16 Ellul, *The Technological Society*, xxxiii.

17 Lewin, *Field Theory in Social Science*, 169.

1. The Network as Method for Organizing the World

1 In order to elaborate on attention capital, a basic understanding of *attention economics* is needed. Attention is "the action that turns raw data into something humans can use" (Lanham, quoted in Lankshear and Knobel, *New Literacies*, 111). Thus attention economics treats attention as a scarce commodity. What information consumes is "the attention of its recipients. Hence a wealth of information creates a poverty of attention and a need to allocate attention efficiently among the overabundance of information sources that might consume it" (Simon, quoted in Lankshear and Knobel, *New Literacies*, 109).

2 Giannone, "World's Rich Got Richer amid '09 Recession."

3 Hardt and Negri, *Empire*, 295.

4 Andrejevic, *iSpy*, 3.

5 Ibid.

6 Smith, "Mobile Access 2010." The same survey found that minority cell phone owners were taking advantage of more of their phone features compared to white mobile phone users.

7 See "Harper's Index," 11; Allegretto, *The State of Working America's Wealth*, 10.

8 Frank, *Capitalism and Underdevelopment in Latin America*.

9 Andrejevic, *iSpy*, 7–8.

10 Mejias, "Between Google and a Hard Place."

11 Details are hard to glean for those not privy to the negotiations and contracts, but apparently, due to the Family Educational Rights and Privacy Act (FERPA), the content of students' e-mails will not be mined for data while they are enrolled, although Google can still track the "signaling data" (web links within e-mails, activity between people, etc.). If students wish to keep their Gmail accounts after they graduate, then their e-mail contents will presumably be open for mining.

12 Gramsci, *Selections from the Prison Notebooks of Antonio Gramsci*.

13 Chatterjee, *The Nation and Its Fragments*.

14 Castells, *The Rise of the Network Society*, 501.

15 This is reminiscent of the old colonialist model of cultural diffusion (cf. Blaut, *The Colonizer's Model of the World*) and more contemporary theorizings of a normal inside and a queer outside (cf. Fuss, *Inside/Out*).

16 Hardt and Negri, *Empire*.

17 Ibid., 211.

18 Balibar, *Masses, Classes, Ideas*, 29.

19 Anderson, *Imagined Communities*.

20 Chatterjee, *The Nation and Its Fragments*.

21 Ibid.

22 Benjamin, "The Work of Art."

23 It is also interesting to note the role of war in occupying the masses according to Benjamin, for "[w]ar and war only can set a goal for mass movements on the largest scale while respecting the traditional property system." Benjamin, "The Work of Art," 241.

2. The Privatization of Social Life

1 It should be noted that Benkler's argument is more nuanced than others that present the collaborative and peer-to-peer models of production and sharing that the Internet makes possible as the most important challenges to capitalism in modern times. See for instance Kelly, "The New Socialism."

2 Ibid., 7.

3 DeTar, "Bike Maps."

4 O'Connor, "Google Maps Finally Adds Bike Routes."

5 Dean, *Democracy and Other Neoliberal Fantasies*, 2.

6 Boltanski and Chiapello, *The New Spirit of Capitalism.*

7 Deleuze, *Negotiations 1972–1990*, 129.

8 Jenkins et al., *Confronting the Challenges of Participatory Culture.* In a participatory culture, digital networks are believed to empower people to become producers and not merely consumers of culture and to become more actively engaged in civic processes.

9 Dean, *Democracy and Other Neoliberal Fantasies.*

10 Dijk, *The Network Society.*

11 Galloway, *Protocol*, 50. The author argues that "at the same time that it is distributed and omnidirectional, the digital network is hegemonic by nature; that is, digital networks are structured on a negotiated dominance of certain flows over other flows" (75).

12 Kücklich, "Michael Jackson and the Death of Macrofame." Kücklich's defines playbor as the "Taylorization of leisure." According to him, this "affective or immaterial" form of labor "is not productive in the sense of resulting in a product." Rather, the process of participation itself generates value. "The means of production are the players themselves, but insofar as they only exist within play environments by virtue of their representations, and their representations are usually owned by the providers of these environments, the players cannot be said to be fully in control of these means."

13 Diken and Laustsen, "Enjoy Your Fight!," 9.

14 Grimes and Feenberg, "Rationalizing Play."

15 Foucault, *Discipline and Punish.*

16 Deleuze, *Negotiations 1972–1990.*

17 Vandenberghe, "Reconstructing Humants."

18 Levine, "A Public Voice for Youth."

19 Montgomery, "Youth and Digital Democracy."

20 Ibid., 42.

21 Coleman, "Doing IT for Themselves."

22 Cox and Knahl, "Critique of Software Security."

23 Carr, "Is Google Making Us Stupid?"

24 For a more extensive discussion of the characteristics and dynamics of social networking websites, see Boyd and Ellison, "Social Network Sites."

25 Britton and McGonegal, *The Digital Economy Fact Book*, 80.

26 Ibid.

27 Eskelsen, Marcus, and Ferree, *The Digital Economy Fact Book*, 102–3.

28 Smith and Rainie, *The Internet and the 2008 Election*, 8.

29 Ibid.

30 Ibid., 10.

31 "Social Networks/Blogs Now Account for One in Every Four and a Half Minutes Online."

32 Rosen, "The People Formerly Known as the Audience."

33 Pasquinelli, *Animal Spirits*.

34 Kuttner, *Everything for Sale*.

35 Nest, *Coltan*.

36 Moore, "Inside Foxconn's Suicide Factory."

37 Yang et al., 2011.

38 Gunther, "News Corp. (hearts) MySpace."

39 Bagdikian, *The New Media Monopoly*; Croteau and Hoynes, *The Business of Media*; Wu, *The Master Switch*.

3. Computers as Socializing Tools

1 Morgan, *Images of Organization*. Morgan does talk about networks at various points in his book, but not as one of his archetypal metaphors for describing organizations.

2 Wellman, *Networks in the Global Village*.

3 Galloway, *Protocol*.

4 Agre, "The Practical Republic," 201–23, para. 22.

5 Michael Mahoney, "The Histories of Computing(s)," 53, in this vein asks, "In seeking to do things in new ways with a computer, it is useful to clarify how we do them now and how we came to do them that way and not otherwise."

6 Dourish, *Where the Action Is*, 1.

7 Campbell-Kelly and Aspray, *Computer*, xv.

8 Tangney and Lytle, "Preface," v.

9 Lanier, "Digital Maoism," para. 4.

10 Scholz, "What the MySpace Generation Should Know about Working for Free."

11 Monge and Contractor, *Theories of Communication Networks*.

12 Barabási, *Linked*.

13 Ibid.

14 Monge and Contractor, *Theories of Communication Networks*.

15 Wellman, "Little Boxes," 11–12.

16 Wellman, "Networks in the Global Village," 2.

17 Van Dijk, *The Network Society*.

18 Wellman, "Networks in the Global Village," 3.

19 See for instance Bourdieu, *Distinction*; Coleman, *Foundations of Social Theory*; Putnam, *Bowling Alone*; Lin, *Social Capital*.

20 Monge and Contractor, *Theories of Communication Networks*, 88.

21 DeLanda, *Intensive Science and Virtual Philosophy*, in this vein advocates for a

"reversal of the problem-solution relation," insofar as "subordinating problems to solutions may be seen as a practice that effectively hides the virtual."

22 Ibid.

23 Committee on Network Science for Future Army Applications, *Network Science*, 28.

24 Ibid.

25 A *collaborative filtering algorithm* mines a large dataset looking for other users who have expressed similar interests or looked at the same objects one has. It then proceeds to identify things they have liked that one has not accessed and puts together a list of suggestions ranked according to which items come from users with closely related interests. A *naïve Bayes classifier* is an algorithm that is optimal for highlighting the things one is interested in or eliminating the things one finds uninteresting. It works by parsing a file and assigning a probability value to each element (such as a word) within it. This probability value represents the likelihood that one is, or is not, interested in that single element. Then the algorithm calculates an overall likelihood that you would or would not be interested in the document as a whole, and highlights or obscures it accordingly. A *decision-tree classifier* works by recursively partitioning a dataset until each partition contains examples from one class only, and it is ideal for organizing things into unique categories. To provide a simplified example, if a user is performing a query for pets that are not only cute but also kid friendly, the system could provide results according to the pets users have ranked as most cute and most kid friendly. Thus finding a match is achieved by eliminating those objects that do not fit the parameters established by the aggregated opinion of users. Lastly, a *k-nearest neighbor* algorithm includes or excludes objects based on the quantity of other objects of a known class found in the proximity. For instance, if there are more blue objects than red objects in the vicinity, the object in question is classified as blue and inferred to have characteristics in common with blue objects. At the same time, the algorithm can be adjusted to consider various degrees of proximity (referred to as k). For instance, if k is broadened (which basically means extending the range of what is considered to be proximal, or how extensively the algorithm is told to look) and a new calculation results in more red objects found in the vicinity than blue objects, the object is reclassified as red. For more information on any of these algorithms, see Segaran, *Programming Collective Intelligence*.

26 According to Dave Boothroyd, "The Ends of Censorship," para. 22, "[C]ensorship could become, perhaps already is becoming, an internal feature and control mechanism of socio-technological systems of governance."

27 Shirky, "Folksonomy."

28 Shirky, "Matt Locke on Folksonomies," para. 10.

29 Locke, "The Politics of the Playful Web," para. 2.

4. Acting Inside and Outside the Network

1 Borgmann, "Is the Internet the Solution to the Problem of Community?," 64.

2 Ibid., 63.

3 Ibid., 65.

4 Ibid., 77.

5 Ibid., 78.

6 Ibid.

7 Dreyfus, "Telepistemology," 49.

8 According to Dreyfus, pragmatics, existential phenomenologists, and language philosophers (such as Dewey, Heidegger, and Wittgenstein) eventually began to question the notion of a brain external to the world and with indirect access to reality. Instead, they proposed that we make sense of the world as a result of being embedded right in it, inseparable from it. "[A]ttempting to prove that there is an external world presupposes a separation of the mind from the world of things and other people which defies a phenomenological description of how human beings make sense of everyday things and of themselves," Dreyfus, "Telepistemology," 53.

9 Walther, "Computer-Mediated Communication."

10 Ibid., 32.

11 Walter Ong, for instance, writes that "electronic technology has brought us into the age of 'secondary orality,'" which resembled the older oral forms in "its participatory mystique, its fostering of a communal sense, its concentration on the present moment" At the same time, Ong recognized the importance of literacy in this new form of orality, which is "based permanently on the use of writing and print, which are essential for the manufacture and operation of the equipment and for it use as well." Ong, *Orality and Literacy*, 133–34.

12 Schutz, *The Phenomenology of the Social World*, 102.

13 Ibid., 143. It is true that some social relations that take place in a web-enabled social network can be with consociates and not just with contemporaries (e.g., my wife—a consociate—might also be part of my 'friends' in a social networking website).

14 Rivers, "An Introduction to the Metaphysics of Technology."

15 Ibid., 575.

16 Heidegger, *The Question Concerning Technology, and Other Essays.*

17 Rivers, *Contra Technologiam*, 10.

18 Brey, "Artifacts as Social Agents."

19 Disco, "Back to the Drawing Board," 58.

20 For an overview of ANT, see Callon, "Society in the Making"; Latour, "Where Are the Missing Masses?"

21 Brey, "Artifacts as Social Agents," 62.

22 Ibid., 74.

23 Latour, *Reassembling the Social*, 8.

24 Wise, *Exploring Technology and Social Space*, 58.

25 Brey, "Artifacts as Social Agents," 79.

26 Dourish, *Where the Action Is*, 97.

27 See for example, Wellman, "Networks in the Global Village"; Dijk, *The Network Society*; Castells, *The Rise of the Network Society*; Castells, "Why Networks Matter."

28 Tocqueville, *Democracy in America*; Dewey, *The Public and Its Problems*;

Lippmann, *The Phantom Public*; Mills, *The Power Elite*; Habermas, *The Structural Transformation.*

29 Mills, *The Power Elite*, 303–4.

30 Ibid.

31 Deleuze, *Negotiations 1972–1990*, 129.

32 Mills, *The Power Elite*, 303–4.

33 Rivers, "An Introduction to the Metaphysics of Technology."

34 Hobbes, *Leviathan*, xiii.

35 Hardt and Negri, *Empire*, 104.

36 Virno, *A Grammar of the Multitude*, xiv.

37 Ibid., 21.

38 Ibid., 63.

39 Grygiel, "The Power of Statelessness," para. 2.

40 Agamben, *Coming Community*, 86.

41 Hardt and Negri, *Empire*, 70.

42 See Colombres, *La Hora del Barbaro*; Sardar, Nandy, and Davies, *Barbaric Others.*

43 Lee and Stenner, "Who Pays?," 105.

5. Strategies for Disrupting Networks

1 See Thierer and Eskelsen, *Media Metrics*, 18, exhibit 4.

2 Terranova, *Network Culture*; Lovink, *Zero Comments*; Dean, *Democracy and Other Neoliberal Fantasies*; Rossiter, *Organized Networks*; Andrejevic, *iSpy*; Morozov, *The Net Delusion*; Hands, *@ Is For Activism*; Galloway and Thacker, *The Exploit.*

3 Hardt and Negri, *Multitude*, 58.

4 Latour, *Reassembling the Social*, 207.

5 Baudrillard, *Simulacra and Simulation.*

6 DeLanda, *Intensive Science and Virtual Philosophy*, 4.

7 Ibid.

8 Colebrook, *Gilles Deleuze*, 71.

9 DeLanda, *Intensive Science and Virtual Philosophy*, 56.

10 Deleuze, *Difference and Repetition*, 208.

11 Ibid., 211.

12 DeLanda, *Intensive Science and Virtual Philosophy*, 6.

13 Deleuze, *Negotiations 1972–1990*, 95.

14 Deleuze, *Difference and Repetition*, xix.

15 Deleuze, "Bergson's Conception of Difference," 53.

16 Ibid.

17 Deleuze, *Cinema* 2, 82–83.

18 DeLanda, *Intensive Science and Virtual Philosophy*, 106.

19 Rivers, *Contra Technologiam*, 106.

20 Ansell-Pearson, *Philosophy and the Adventure of the Virtual*, 184.

21 Colebrook, *Gilles Deleuze*, 13.

22 See Serres, *The Parasite*; Crocker, "Noises and Exceptions."

23 Lyotard, *The Postmodern Condition.*

24 Gane, "Computerized Capitalism," 438.

25 Readings, *Introducing Lyotard*, 73–74.

26 I am taking a cue here from McKenzie Wark, who discusses some aspects of video game narrative in terms of utopia, heterotopia, and atopia. Wark, *Gamer Theory*.

27 From the Greek *ou* or *not* and *tóp* or *place*.

28 From the Greek *héteros* or *different*. For more on heterotopias, see Foucault, "Of Other Spaces."

29 Here, the prefix *a-* suggests without.

30 Virno, *A Grammar of the Multitude*, 70.

31 Winner, *Autonomous Technology*, 315.

32 Latour and Venn, "Morality and Technology," 258.

33 Ibid., 257.

6. Proximity and Conflict

1 Boo, "The Best Job in Town," 40.

2 Ibid., 65.

3 Ibid., 59.

4 Eliot, "Why the Web Won't Ruin the World," para. 9.

5 Ibid., para. 16.

6 Silverstone, *Why Study the Media?*, 151.

7 Heidegger, *Poetry, Language, Thought*, 164.

8 Borgmann, "Information, Nearness, and Farness," 98.

9 Latour, *We Have Never Been Modern*.

10 Yus, "The Linguistic-Cognitive Essence of Virtual Community," 87.

11 This emphasis on the functionality of knowledge "is traceable in its lineage to the popular belief . . . that tacit knowledge can be converted into explicit knowledge through IT systems. By capturing knowledge, it can be more widely replicated and shared. . . . Henceforth, knowledge is transformed into a more tangible commodity." Chan and Garrick, "The Moral Technologies of Knowledge Management," 291.

12 Galloway, "Intimations of Everyday Life," 397.

13 Sassi, "Cultural Differentiation or Social Segregation?"

14 Massey, *Power-Geometries and the Politics of Space-Time*, cited in Rodgers, "Doreen Massey," 287.

15 Lyotard, *The Inhuman*, 64.

16 Castells, *The Rise of the Network Society*.

17 See Arquilla and Ronfeldt, *Networks and Netwars*; Galloway and Thacker, *The Exploit*.

18 Hass, "Holding on Tight to the Frequencies."

19 See Carr, *Inside Cyber Warfare*; Andress and Winterfeld, *Cyber Warfare*.

20 Hardt and Negri, *Empire*; Hardt and Negri, *Multitude*.

21 One website related to the campaign is http://zyprexakills.ath.cx.

22 Sliva et al., "The SOMA Terror Organization Portal (STOP)."

23 McCahill and Norris, *On the Threshold to Urban Panopticon?*

24 Klein, "China's All-Seeing Eye."
25 Wilkinson, "Non-Lethal Force."
26 Abelson, Ledeen, and Lewis, *Blown to Bits*.
27 For a detailed discussion of some of these strategies, see Rheingold, *Smart Mobs*; Gillmor, *We the Media*; Burgess and Green, *YouTube*; Shirky, *Here Comes Everybody*.
28 Lih, "What Does Cyber-Revolt Look Like?," para. 39.
29 Zuckerman, "Does the Number Have a Lesson for Human Rights Activists?," para. 12.
30 *Bloomberg Businessweek*, "Valley Boys."
31 The author was Ryan Shaw. The blog post is no longer available, but I decided to keep the quote because it is quite powerful, in my opinion.
32 Dreyfus, "Nihilism on the Information Highway."
33 Deleuze, *Negotiations 1972–1990*, 129.
34 Shapiro, "Revolution, Facebook-Style," para. 44.
35 Shapiro, "Revolution, Facebook-Style."
36 Ibid.
37 Norman and York, "West Censoring East."
38 York, "This Week in Internet Censorship"; Mayton, "U.S. Company May Have Helped Egypt Spy on Citizens."
39 Abbott, "Torture Victims Say Cisco Systems Helped China Hound and Surveil."
40 MacKinnon, "Internet Freedom."
41 Glanz and Markoff, "U.S. Underwrites Internet Detour around Censors Abroad."
42 Fielding and Cobain, "Revealed."
43 York, "Syria's Twitter Spambots."
44 Malik, "Facebook Accused of Removing Activists' Pages."
45 Shenker, "Fury over Advert Claiming Egypt Revolution as Vodafone's."
46 McCullagh and Broache, "FBI Taps Cell Phone Mic as Eavesdropping Tool."
47 Mack, "Patent Application Suggests Infrared Sensors for iPhone."
48 Wong, "Social Media Play Big Role in Riot Probe."
49 Tehrani, "Iranian Officials 'Crowd-Source' Protester Identities."
50 Morozov, "Testimony to the U.S. Commission on Security and Cooperation in Europe"; Morozov, *The Net Delusion*.
51 Khatri, "Facebook Usage Falls in GCC, Including in Qatar, Saudi Arabia."
52 Fuchs, *Social Networking Sites and the Surveillance Society*, 115.
53 Shapiro, "Revolution, Facebook-Style," para. 29.
54 *Wikipedia*, "Neutral Point of View."
55 Rancière, *On the Shores of Politics*, 103.
56 Virno, *A Grammar of the Multitude*.
57 Rancière, *On the Shores of Politics*, 104.
58 Ibid.
59 Kothari and Mehta, "Cancer."
60 Galloway, *Protocol*.
61 Sützl, "Tragic Extremes," para. 2.
62 Barlas, *"Believing Women" in Islam*.

63 Heneghan, "French Muslim Council Warns Government on Veil Ban"; Heneghan, "Barcelona to Ban Veil in Public Buildings."

64 Solomon, *War Made Easy*.

65 Singer, *Wired for War*, 268.

66 Serrie, "Propaganda War Rages Online."

67 Morozov, "An Army of Ones and Zeroes."

68 Robb, "Open Source Warfare."

69 Organizations like the Russian Business Network are experts in mounting Denial of Service attacks for extortion. See "RBN—Extortion and Denial of Service (DDOS) Attacks."

70 Singer, *Wired for War*, 179, 184.

71 Ibid., 348.

72 Bronner and Richards, "Integrating Multi-Agent Technology," 28.

73 Sliva et al., "The SOMA Terror Organization Portal (STOP)."

74 Ibid., 39.

75 Ibid.

76 Ibid., 12.

77 Mannes et al., "Stochastic Opponent Modeling Agents."

78 Bronner and Richards, "Integrating Multi-Agent Technology," 29.

7. Collaboration and Freedom

1 *Commons-based peer production* and *social production* are terms used by Benkler. *Wikinomics* is a term defined by Tapscott and Williams. See Benkler, *The Wealth of Networks*; Tapscott and Williams, *Wikinomics*.

2 According to Google's website (YouTube is owned by Google): "If a rights owner specifies a Block policy, the video will not be viewable on YouTube. If the rights owner specifies a Track policy, the video will continue to be made available on You-Tube and the rights owner will receive information about the video, such as how many views it receives. For a Monetize policy, the video will continue to be available on YouTube and ads will appear in conjunction with the video. The policies can be region-specific, so a content owner can allow a particular piece of material in one country and block the material in another." See http://www.google.com/support/youtube/bin/answer.py?hl=en&answer=83766.

3 Doctorow, "Record Company Embraces Use of Its Music."

4 An Internet meme is a concept, object, or hyperlink that spreads rapidly from person to person by means of digital networks (metaphorically, a *meme* copies and distributes its information similarly to a *gene*, but the information consists of ideas and not genetic material, and its hosts are brains and not cells).

5 For some popular examples of this parody, see http://www.telegraph.co.uk/technology/news/6262709/Hitler-Downfall-parodies-25-worth-watching.html.

6 See http://www.youtube.com/watch?v=kBO5dh9qrIQ.

7 Ibid.

8 P2P Foundation, "What This Essay Is About."

9 Bauwens, "The Political Economy of Peer Production," para. 55.

10 Foucault, "Of Other Spaces."

11 Pasquinelli, *Animal Spirits*, 66

12 Ibid., 72–73; emphasis in original.

13 Kelly, "The New Socialism."

14 Virno, *A Grammar of the Multitude*, 18.

15 Ibid., 110.

16 Hardt and Negri, *Empire*, 385.

17 Virno, *A Grammar of the Multitude*, 18.

18 Ibid., 40.

19 According to some figures, 750,000 jobs and up to $250 billion are lost every year because of piracy, although many of the estimates surrounding piracy have been contested, even by the government. See Anderson, "U.S. Government Finally Admits Most Piracy Estimates Are Bogus."

20 Dyer-Witheford and de Peuter, "Empire@Play," para. 28.

21 Business Software Alliance/Harris Interactive, *Youth and Downloading Behavior*.

22 *IFPI.org*, "IFPI Publishes Digital Music Report 2010."

23 Some industry leaders like Irving Azoff believe that stealing music actually generates more interest in concerts, which is where most pop stars are making their money today. While an artist like Bruce Springsteen might only make $10 million in record sales, he might make as much as $200 million on a single tour. See Seabrook, "The Price of the Ticket."

24 Searls, "Power Re-Origination," para. 14.

25 Richburg, "Google Compromise Pays Off with Renewal of License in China."

26 See, for instance, "Authority, Meet Technology."

27 Clinton, "Remarks on Internet Freedom."

28 Ibid., para. 17.

29 Ibid., para. 52.

30 Machet, "U.S. Congress Bill Threatens."

31 See Lynch, "Arabs Reject U.S. Crackdown," para. 4. Lynch laments the disjunction between this bill and Clinton's vision for an open media: "H.R. 2278 is a deeply irresponsible bill which sharply contradicts American support for media freedom and could not be implemented in the Middle East today as crafted without causing great damage . . . The last thing the Arab world needs right now is more state power of censorship over the media—whether the Arab League over satellite TV or the Jordanian government over the internet. Hillary Clinton just laid out a vision of an America committed to internet freedom, and that should be embraced as part of a broader commitment to free and open media. Nobody should be keen on restoring the power of authoritarian governments over one of the few zones of relative freedom which have evolved over the last decade" (para. 5).

32 Clinton, "Remarks on Internet Freedom," para. 48.

33 Rhoads and Chao, "Iran's Web Spying Aided by Western Technology."

34 Clinton, "Remarks on Internet Freedom," para. 55.

35 Ibid., para. 34.

36 Howcast.com, "Alliance of Youth Movements," para. 1, 2.

37 Lichtenstein, "Digital Diplomacy."

38 Ibid.

39 Clinton, "Remarks on Internet Freedom," para. 26.

8. The Limits of Liberation Technologies

1 Abu-Samra, "Expulsion and Explosion," para. 3, 9, 10.

2 For a discussion of these and other examples, see Hands, @ *Is For Activism*; Rheingold, *Smart Mobs*; Morozov, *The Net Delusion*.

3 The website for the program is http://liberationtechnology.stanford.edu.

4 Dean, *Democracy and Other Neoliberal Fantasies*.

5 World without Oil, homepage, http://worldwithoutoil.org.

6 For an overview of PAR, see Fals-Borda and Rahman, *Action and Knowledge*; Cahill, "Including Excluded Perspectives in Participatory Action Research."

7 Save Oswego, About, http://saveoswego.wordpress.com/about.

8 For a more detailed discussion of the collective writing process in the ARG, see Clark et al., "Interactive Social Media and the Art of Telling Stories."

9 This is a feature of the ARG we started implementing in the second year, and not something that is a standard practice across all serious ARG, although some games such as *Urgent Evoke* have implemented other innovative ways of "going beyond" the digital network.

9. The Outside of Networks as a Method for Acting in the World

1 Rancière, *Disagreement*.

2 Lovink asks, "What is linking and how could we describe its mirror phantom?" Lovink, *Zero Comments*, 235.

3 Wyatt, Thomas, and Terranova, "They Came, They Surfed."

4 Lee and Stenner, "Who Pays?," 105.

5 Barabási, *Linked*.

6 Hoy, *Critical Resistance*.

7 The story is "Del rigor en la ciencia" ("On Exactitude in Science").

8 Coleman, "Anonymous," para. 15.

9 Winner, *Autonomous Technology*.

10 This entropy is captured in what I call the bang-boost-burst-and-purge life cycle of digital networks: Bang: an initial period of rapid growth, as early adopters rush to join the hot new app; emergence of rich nodes (which will become richer through preferential attachment). Boost: a period of capitalization; investment accelerates growth, and the network achieves critical mass, as the inequality between rich nodes and poor nodes is converted into wealth for investors. Burst: a period when hyperinflation leads to bubble popping. Purge: in the aftermath of the crisis, investors reap the rewards, while users loose their content (their wealth) or are forced to accept new terms of use; unwanted nodes and modes of participation (fake profiles, for instance) are purged from the network.

11 Andrejevic, *iSpy*, 243

12 For an example of one such plea, see for instance http://theithacan.org/5564.

13 Amazon refused to host WikiLeaks's website; Bank of America stopped processing contributions to the organization; and Apple removed the WikiLeaks iPhone app from its market.

BIBLIOGRAPHY

Abbott, Ryan. "Torture Victims Say Cisco Systems Helped China Hound and Surveil." *Courthouse News Service*, June 10, 2011. http://www.courthousenews .com/2011/06/10/37266.htm.

Abelson, Harold, Ken Ledeen, and Harry R Lewis. *Blown to Bits: Your Life, Liberty, and Happiness after the Digital Explosion*. Upper Saddle River, N.J.: Addison-Wesley, 2008.

Abu-Samra, Haisam. "Expulsion and Explosion: How Leaving the Internet Fueled Our Revolution." *Mother Board*, February 3, 2011. http://www.motherboard.tv.

Agamben, Giorgio. *Coming Community*. 1st ed. University of Minnesota Press, 1993.

Allegretto, Sylvia. *The State of Working America's Wealth, 2011*. Economic Policy Institute, March 24, 2011. http://www.epi.org/publications/.

"Alliance of Youth Movements to Convene Second-Annual Summit on Social Change and the Role of Technology." *Howcast.com*, October 1, 2009. http://info.howcast .com/youthmovements/summit09/press/pressrelease.

Anderson, Benedict R. O'G. *Imagined Communities: Reflections on the Origin and Spread of Nationalism*. London: Verso, 1983.

Anderson, Nate. "U.S. Government Finally Admits Most Piracy Estimates Are Bogus." *Ars Technica*, April 2010. http://arstechnica.com.

Andrejevic, Mark. *iSpy: Surveillance and Power in the Interactive Era*. University Press of Kansas, 2009.

Andress, Jason, and Steve Winterfeld. *Cyber Warfare: Techniques, Tactics and Tools for Security Practitioners*. Syngress, 2011.

Ansell-Pearson, Keith. *Philosophy and the Adventure of the Virtual*. Routledge, 2001.

Arquilla, John, and David F. Ronfeldt, eds. *Networks and Netwars: The Future of Terror, Crime, and Militancy*. Santa Monica, Calif.: Rand, 2001. http://www.rand.org/ pubs/monograph_reports/MR1382/index.html.

"Authority, Meet Technology." *Slate.com*, January 20, 2010. http://www.slate.com/ id/2241755/workarea/3/.

Bagdikian, Ben H. *The New Media Monopoly*. Beacon Press, 2004.

Baldwin, James. *No Name in the Street*. Vintage, 2007.

Balibar, Etienne. *Masses, Classes, Ideas: Studies on Politics and Philosophy before and after Marx*. New York: Routledge, 1994.

Barabási, Albert-László. *Linked: The New Science of Networks*. Basic Books, 2003.

"Barcelona to Ban Veil in Public Buildings." *Reuters*, June 14, 2010. http://www.reuters.com/article/idUSTRE65D4K620100614.

Barlas, Asma. *"Believing Women" in Islam: Unreading Patriarchal Interpretations of the Qur'an*. University of Texas Press, 2002.

Baudrillard, Jean. *Simulacra and Simulation*. Translated by Sheila Faria Glaser. University of Michigan Press, 1994.

Bauwens, Michel. "The Political Economy of Peer Production." *CTheory.net*, December 1, 2005. http://www.ctheory.net/articles.aspx?id=499.

Benjamin, Walter. "The Work of Art in the Age of Mechanical Reproduction." In *Illuminations: Essays and Reflections*, edited by Hannah Arendt, translated by Harry Zohn, 217–51. New York: Harcourt and Brace, 1968.

Benkler, Yochai. *The Wealth of Networks: How Social Production Transforms Markets and Freedom*. New Haven, Conn.: Yale University Press, 2006.

Blaut, James M. *The Colonizer's Model of the World: Geographical Diffusionism and Eurocentric History*. New York: Guilford Press, 1993.

Boltanski, Luc, and Eve Chiapello. *The New Spirit of Capitalism*. Verso, 2007.

Bonk, Curtis Jay. *The World Is Open: How Web Technology Is Revolutionizing Education*. John Wiley and Sons, 2009.

Boo, Katherine. "The Best Job in Town." *The New Yorker*, July 5, 2004.

Boothroyd, Dave. "The Ends of Censorship." *Eurozine*, May 26, 2009. http://www.eurozine.com/articles/2009-05-26-boothroyd-en.html.

Borgmann, Albert. "Information, Nearness, and Farness." In *The Robot in the Garden: Telerobotics and Telepistemology in the Age of the*, edited by Ken Goldberg, 90–107. Cambridge, Mass.: MIT Press, 2000.

Bossewitch, Jonah. "The Tweets of War." *Alchemical Musings*, May 6, 2006. http://alchemicalmusings.org/2009/01/19/the-tweets-of-war/.

Bourdieu, Pierre. *Distinction: A Social Critique of the Judgement of Taste*. Cambridge, Mass.: Harvard University Press, 1984.

Boyd, Danah M., and Nicole B. Ellison. "Social Network Sites: Definition, History, and Scholarship." *Journal of Computer-Mediated Communication* 13, no. 1 (2007): 210–30. http://jcmc.indiana.edu/vol13/issue1/boyd.ellison.html.

Brey, Philip. "Artifacts as Social Agents." In *Inside the Politics of Technology: Agency and Normativity in the Co-Production of Technology and Society*, edited by Hans Harbers, 61–84. Amsterdam: Amsterdam University Press, 2005.

Britton, Daniel B., and Stephen McGonegal. *The Digital Economy Fact Book, Ninth Edition*. Washington, D.C.: The Progress & Freedom Foundation, 2007.

Bronner, LeeRoy, and Akeila Richards. "Integrating Multi-Agent Technology with Cognitive Modeling to Develop an Insurgency Information Framework (IIF)." In *Social Computing, Behavioral Modeling, and Prediction*, edited by Huan Liu, John Salerno, and Michael J. Young, 26–36. New York: Springer, 2008.

Burgess, Jean, and Joshua Green. *YouTube: Online Video and Participatory Culture*. Polity, 2009.

Cahill, C. "Including Excluded Perspectives in Participatory Action Research." *Design Studies* 28 (2007): 325–40.

Callon, Michael. "Society in the Making: The Study of Technology as a Tool for Sociological Analysis." In *The Social Construction of Technological Systems: New Directions in the Sociology and History of Technology*, edited by Wiebe Bijker, Thomas P. Hughes, and Trevor Pinch, 83–103. Cambridge, Mass.: MIT Press, 1987.

Campbell-Kelly, Martin, and William Aspray. *Computer: A History of the Information Machine.* 2nd ed. Westview Press, 2004.

Carr, Jeffrey. *Inside Cyber Warfare: Mapping the Cyber Underworld.* O'Reilly Media, 2009.

Carr, Nicholas. "Is Google Making Us Stupid?" *Atlantic Monthly*, August 2008. http://www.theatlantic.com/doc/200807/google.

Castells, Manuel. *The Rise of the Network Society.* 2nd ed. Oxford: Blackwell Publishers, 2000.

———. "Why Networks Matter." In *Network Logic: Who Governs in an Interconnected World?*, edited by H. McCarthy, P. Miller, and P. Skidmore, 219–25. London: Demos, 2004. http://www.demos.co.uk/publications/networks.

Chan, Andrew, and John Garrick. "The Moral Technologies of Knowledge Management." *Information, Communication and Society* 6 (September 2003): 291–306.

Chatterjee, Partha. *The Nation and Its Fragments: Colonial and Postcolonial Histories.* Princeton Studies in Culture/Power/History. Princeton, N.J.: Princeton University Press, 1993.

Clark, Patricia E., Ulises A. Mejias, Peter Cavana, Daniel Herson, and Sharon M. Strong. "Interactive Social Media and the Art of Telling Stories: Strategies for Social Justice through Osw3go.net 'Racism on Campus'." In *Social Justice through the Arts*, edited by Barbara Beyerbach and R. Deborah Davis, 171–85. New York: Peter Lang, 2011.

Colebrook, Claire. *Gilles Deleuze.* 2nd ed. Routledge, 2012.

Coleman, E. Gabriella. "Anonymous: From the Lulz to Collective Action." *The New Everyday: A Media Commons Project*, April 6, 2011. http://mediacommons.futureofthebook.org/.

Coleman, James Samuel. *Foundations of Social Theory.* Cambridge, Mass.: Belknap Press of Harvard University Press, 1990.

Coleman, Stephen. "Doing IT for Themselves: Management versus Autonomy in Youth E-Citizenship." *The John D. and Catherine T. MacArthur Foundation Series on Digital Media and Learning* (December 1, 2007): 189–206. http://www.mitpressjournals.org.

Colombres, A. *La Hora del Barbaro.* Buenos Aires, Argentina: Ediciones del Sol, 1988.

Committee on Network Science for Future Army Applications. *Network Science.* Washington, D.C.: The National Academies Press, 2005. http://www.nap.edu/catalog.php?record_id=11516.

Cox, Geoff, and Martin Knahl. "Critique of Software Security." In *Creating Insecurity: Art and Culture in the Age of Security*, edited by Wolfgang Sützl and Geoff Cox, 27–43. New York: Autonomedia, 2009. http://www.anti-thesis.net/contents/texts/violence.pdf.

Crocker, Stephen. "Noises and Exceptions: Pure Mediality in Serres and Agamben." *cTheory*, March 28, 2007. http://www.ctheory.net/articles.aspx?id=574.

Croteau, David R., and William Hoynes. *The Business of Media: Corporate Media and the Public Interest*. 2nd ed. Thousand Oaks, Calif.: Pine Forge, 2005.

Dean, Jodi. *Democracy and Other Neoliberal Fantasies: Communicative Capitalism and Left Politics*. Durham, N.C.: Duke University Press, 2009.

DeLanda, Manuel. *Intensive Science and Virtual Philosophy*. Continuum International Publishing Group, 2005.

Deleuze, Gilles. "Bergson's Conception of Difference." In *The New Bergson*, edited by John Mullarkey, translated by McMahon Melissa, 43–65. Manchester, U.K.: Manchester University Press, 1999.

———. *Cinema 2: The Time-Image*. University Of Minnesota Press, 1989.

———. *Difference and Repetition*. New York: Columbia University Press, 1994.

———. *Negotiations 1972–1990*. New York: Columbia University Press, 1997.

Deleuze, Gilles, and Félix Guattari. *A Thousand Plateaus: Capitalism and Schizophrenia*. Continuum, 2001.

DeTar, Charlie. "Bike Maps: Triumph of Corporate Solutions Over Grassroots?" *MIT Center for Civic Media*, March 10, 2010. http://civic.mit.edu/blog/cfd/bike-maps-triumph-of-corporate-solutions-over-grassroots.

Dewey, John. *The Public and Its Problems*. Athens: Swallow Press, 1991.

Dijk, Jan A. G. M. van. *The Network Society: Social Aspects of New Media*. London: Sage, 1999.

Diken, Bülent, and Carsten Bagge Laustsen. "Enjoy Your Fight!—'Fight Club' as a Symptom of the Network Society." *Journal for Cultural Research* 6, no. 4 (2002): 349–67. http://www.informaworld.com/10.1080/1362517022000047307.

Disco, Cornelis. "Back to the Drawing Board: Inventing a Sociology of Technology." In *Inside the Politics of Technology: Agency and Normativity in the Co-Production of Technology and Society*, edited by Hans Harbers, 29–60. Amsterdam: Amsterdam University Press, 2005.

Doctorow, Cory. "Record Company Embraces Use of Its Music in YouTube Wedding Video, Makes Money." *BoingBoing.net*, July 31, 2009. http://www.boingboing.net/2009/07/31/record-company-embra.html.

Dourish, Paul. *Where the Action Is: The Foundations of Embodied Interaction*. MIT Press, 2004.

Dreyfus, Hubert. "Nihilism on the Information Highway: Anonymity versus Commitment in the Present Age." In *Community in the Digital Age: Philosophy and Practice*, edited by Darin Barney and Andrew Feenberg, 69–82. Lanham, Md.: Rowman & Littlefield, 2004.

———. "Telepistemology: Descartes's Last Stand." In *The Robot in the Garden: Telerobotics and Telepistemology in the Age of the*, edited by Ken Goldberg, 48–63. MIT Press, 2000.

Dyer-Witheford, Nick, and Greig de Peuter. "Empire@Play: Virtual Games and Global Capitalism." *CTheory.net*, May 13, 2009. http://www.ctheory.net/articles.aspx?id=608.

Eliot. "Why the Web Won't Ruin the World." *Red Linked*, October 27, 2006. http://www.redinked.com/ [no longer available].

Ellul, Jacques. *The Technological Society*. Translated by John Wilkinson. Vintage Books, 1964.

Eskelsen, Grant, Adam Marcus, and W. Kenneth Ferree. *The Digital Economy Fact Book, Tenth Edition*. Washington, D.C.: The Progress and Freedom Foundation, 2009.

Fals-Borda, Orlando, and Muhammad Anisur Rahman, eds. *Action and Knowledge: Breaking the Monopoly with Participatory Action Research*. New York: Apex Press, 1991.

Feld, Lowell, and Nate Wilcox. *Netroots Rising: How a Citizen Army of Bloggers and Online Activists Is Changing American Politics*. Greenwood, 2008.

Fielding, Nick, and Ian Cobain. "Revealed: U.S. Spy Operation that Manipulates Social Media." *The Guardian*, March 17, 2011. http://www.guardian.co.uk/.

Foucault, Michel. *Discipline and Punish: The Birth of the Prison*. New York: Vintage Books, 1979.

———. "Of Other Spaces," 1967. http://foucault.info/documents/heteroTopia/foucault.heteroTopia.en.html.

Frank, Andre Gunder. *Capitalism and Underdevelopment in Latin America: Historical Studies of Chile and Brazil*. Monthly Review Press, 1967.

Fuchs, Christian. *Social Networking Sites and the Surveillance Society: A Critical Case Study of the Usage of studiVZ, Facebook, and MySpace by Students in Salzburg in the Context of Electronic Surveillance*. Salzburg, Vienna: ICT&S Center, University of Salzburg, 2009. http://fuchs.icts.sbg.ac.at/SNS_Surveillance_Fuchs.pdf.

Fuss, Diana, ed. *Inside/Out: Lesbian Theories, Gay Theories*. New York: Routledge, 1991.

Galloway, Alexander R. *Protocol: How Control Exists After Decentralization*. Cambridge, Mass.: MIT Press, 2004.

Galloway, Alexander R, and Eugene Thacker. *The Exploit: A Theory of Networks*. Minneapolis: University of Minnesota Press, 2007.

Galloway, Anne. "Intimations of Everyday Life: Ubiquitous Computing and the City." *Cultural Studies* 18, no. 2 (2004): 384–408.

Gane, Nicholas. "Computerized Capitalism: The Media Theory of Jean-Francois Lyotard." *Information Communication Society* 6, no. 3 (2003): 430–50.

Giannone, Joseph A. "World's Rich Got Richer amid '09 Recession: Report." *Reuters*, June 22, 2010. http://www.reuters.com/article/idUSTRE65L36T20100622.

Gillmor, Dan. *We the Media*. O'Reilly Media, 2004.

Glanz, James, and John Markoff. "U.S. Underwrites Internet Detour Around Censors Abroad." *The New York Times*, June 12, 2011, sec. World. http://www.nytimes.com/.

Gramsci, Antonio. *Selections from the Prison Notebooks of Antonio Gramsci*. 1st ed. New York: International Publishers, 1972.

Grimes, Sara, and Andrew Feenberg. "Rationalizing Play: A Critical Theory of Digital Gaming." *The Information Society* 25, no. 2 (March 2009): 105–18.

Grygiel, Jacob. "The Power of Statelessness". Hoover Institution at Stanford University, April 1, 2009. http://www.hoover.org/publications/policy-review/article/5568.

Gunther, Marc. "News Corp. (hearts) MySpace." *CNN.com*, March 29, 2006. http://
money.cnn.com/2006/03/28/technology/pluggedin_fortune/.

Habermas, Jürgen. *The Structural Transformation of the Public Sphere: An Inquiry into
a Category of Bourgeois Society*. Studies in Contemporary German Social Thought.
Cambridge, Mass.: MIT Press, 1991.

Hands, Joss. *@ Is For Activism: Dissent, Resistance and Rebellion in a Digital Culture*.
Pluto Press, 2011.

Hardt, Michael, and Antonio Negri. *Empire*. Cambridge, Mass.: Harvard University
Press, 2000.

———. *Multitude: War and Democracy in the Age of Empire*. New York: Penguin, 2004.

"Harper's Index." *Harper's Magazine*, August 2010.

Hass, Amira. "Holding On Tight to the Frequencies." *Haaretz.com*, May 31, 2005. http://
www.haaretz.com/.

Heidegger, Martin. *Poetry, Language, Thought*. Harper Perennial Modern Classics, 2001.

———. *The Question Concerning Technology, and Other Essays*. Edited by William Lovitt.
Harper & Row, 1977.

Heneghan, Tom. "French Muslim Council Warns Government on Veil Ban." *Reuters*,
June 4, 2010. http://www.reuters.com/.

Hobbes, Thomas. *Leviathan: With Selected Variants from the Latin Edition of 1668*.
Hackett, 1994.

Hoy, David Couzens. *Critical Resistance: From Poststructuralism to Post-Critique*. Cambridge, Mass.: MIT Press, 2004.

"IFPI Publishes Digital Music Report 2010." *IFPI.org*, January 21, 2010. http://www.ifpi
.org/content/section_resources/dmr2010.html.

Jenkins, H., K. Clinton, R. Purushotma, A. J. Robinson, and M. Weigel. *Confronting the
Challenges of Participatory Culture: Media Education for the 21st Century*. Chi-
cago, Ill.: MacArthur Foundation, 2006. Google Scholar. http://digitallearning
.macfound.org/.

Kelly, Kevin. "The New Socialism: Global Collectivist Society Is Coming Online." *Wired
Magazine*, May 22, 2009. http://www.wired.com/.

Khatri, Shabina. "Facebook Usage Falls in GCC, Including in Qatar, Saudi Arabia." *Doha
News*, May 2011. http://dohanews.co/.

Kiss, Jemima. "Facebook: Did Anyone Really Quit?" *The Guardian*, June 1, 2010, sec.
PDA: The Digital Content Blog. http://www.guardian.co.uk/.

Klein, Naomi. "China's All-Seeing Eye." *Naomiklein.org*, May 14, 2008. http://www
.naomiklein.org/articles/2008/05/chinas-all-seeing-eye.

Kothari, Manu, and Lopa Mehta. "Cancer." In *The Future of Knowledge & Culture: A
Dictionary for the 21st Century*, edited by Vinay Lal and Ashis Nandy, 25–30. New
Delhi: Penguin, 2005.

Kücklich, Julian. "Michael Jackson and the Death of Macrofame." *iDC*, June 26, 2007.
https://lists.thing.net/pipermail/idc/2009-June/003664.html.

Kuttner, Robert. *Everything for Sale: The Virtues and Limits of Markets*. New York: Alfred
A. Knopf, 1997.

Lanier, Jaron. "Digital Maoism: The Hazards of the New Online Collectivism." *Edge*, May 30, 2006. http://www.edge.org/documents/archive/edge183.html.

Lankshear, Colin, and Michele Knobel. *New Literacies: Everyday Practices and Classroom Learning*. Philadelphia: Open University Press, 2003.

Latour, Bruno. *Reassembling the Social: An Introduction to Actor-Network-Theory*. Oxford: Oxford University Press, 2005.

———. *We Have Never Been Modern*. Translated by Catherine Porter. Harvard University Press, 1993.

———. "Where Are the Missing Masses? The Sociology of a Few Mundane Artifacts." In *Shaping Technology / Building Society: Studies in Sociotechnical Change*, edited by Wiebe Bijker and John Law, 223–58. Cambridge, Mass.: MIT Press, 1992.

Latour, Bruno, and Couze Venn. "Morality and Technology." *Theory, Culture & Society* 19, no. 5–6 (December 1, 2002): 247–60. http://tcs.sagepub.com/content/19/5-6/247.abstract.

Lee, Nick, and Paul Stenner. "Who Pays? Can We Pay Them Back?" In *Actor Network Theory and After*, edited by John Law and John Hassard, 90–112. Oxford: Blackwell/Sociological Review, 1999.

Levine, Peter. "A Public Voice for Youth: The Audience Problem in Digital Media and Civic Education." *The John D. and Catherine T. MacArthur Foundation Series on Digital Media and Learning* (December 1, 2007): 119–38. http://www.mitpressjournals.org/.

Lewin, Kurt. *Field Theory in Social Science: Selected Theoretical Papers*. 1st ed. New York: Harper, 1951.

Lichtenstein, Jessica. "Digital Diplomacy." *The New York Times*, June 16, 2010. http://www.nytimes.com/.

Lih, Andrew. "What Does Cyber-Revolt Look Like?" *andrewlih.com*, May 2, 2007. http://www.andrewlih.com/.

Lin, Nan. *Social Capital: A Theory of Social Structure and Action*. Cambridge: Cambridge University Press, 2001.

Lippmann, Walter. *The Phantom Public*. New Brunswick, N.J.: Transaction Publishers, 1993.

Locke, Matt. "The Politics of the Playful Web." *Test*, March 3, 2005. http://test.org.uk/2005/03/03/the-politics-of-the-playful-web/.

Lovink, Geert. *Zero Comments: Blogging and Critical Internet Culture*. New York: Routledge, 2007.

Lynch, Mark. "Arabs Reject U.S. Crackdown on Arab Satellite TV." *Foreign Policy*, January 25, 2010. http://lynch.foreignpolicy.com/.

Lyotard, Jean François. *The Postmodern Condition: A Report on Knowledge*. University of Minnesota Press, 1984.

———. *The Inhuman: Reflections on Time*. Stanford University Press, 1991.

Machet, Cari. "U.S. Congress Bill Threatens to Crackdown on Terror TV." *World Focus*, February 8, 2010. http://worldfocus.org/.

Mack, Eric. "Apple Patent Suggests Infrared Sensors for iPhone." *CNet News*, June 2, 2011. http://news.cnet.com/.

MacKinnon, Rebecca. "'Internet Freedom' in the Age of Assange." *Foreign Policy*, February 17, 2011. http://www.foreignpolicy.com/.

Mahoney, Michael. "The Histories of Computing(s)." *Interdisciplinary Science Reviews* 30 (June 1, 2005): 119–35.

Malik, Shiv. "Facebook Accused of Removing Activists' Pages." *The Guardian*, April 29, 2011. http://www.guardian.co.uk/.

Mannes, Aaron, Mary Michael, Amy Pate, Amy Sliva, V.S. Subrahmanian, and Jonathan Wilkenfeld. "Stochastic Opponent Modeling Agents: A Case Study with Hezbollah." In *Social Computing, Behavioral Modeling, and Prediction*, edited by Huan Liu, John Salerno, and Michael J. Young, 37–45. New York: Springer, 2008.

Massey, Doreen. *Power-Geometries and the Politics of Space-Time: Hettner-Lecture 1998*. Heidelberg, Germany: Dept. of Geography, University of Heidelberg, 1999.

Mayton, Joseph. "U.S. Company May Have Helped Egypt Spy on Citizens." *IT News Africa*, February 10, 2011. http://www.itnewsafrica.com/.

McCahill, Michael, and Clive Norris. *On the Threshold to Urban Panopticon? Analysing the Employment of CCTV in European Cities and Assessing Its Social and Political Impacts*. Hull, United Kingdom: Center for Criminology and Criminal Justice, University of Hull, June 2002. http://www.urbaneye.net/results/ue_wp6.pdf.

McCullagh, Declan, and Anne Broache. "FBI Taps Cell Phone Mic as Eavesdropping Tool." *Cnet News*, December 1, 2006. http://news.cnet.com/.

Mejias, Ulises A. "Between Google and a Hard Place." *The Oswegonian*, April 16, 2010, sec. Opinion. http://www.oswegonian.com/.

Micek, Deborah, and Warren Whitlock. *Twitter Revolution*. Xeno Press, 2008.

Mills, C. Wright. *The Power Elite*. New York: Oxford University Press, 1956.

Monge, Peter R, and Noshir S Contractor. *Theories of Communication Networks*. Oxford: Oxford University Press, 2003.

Montgomery, Kathryn C. "Youth and Digital Democracy: Intersections of Practice, Policy, and the Marketplace." *The John D. and Catherine T. MacArthur Foundation Series on Digital Media and Learning* (December 1, 2007): 25–49. http://www.mit pressjournals.org/.

Moore, Malcolm. "Inside Foxconn's Suicide Factory." *The Telegraph*, May 27, 2010. http://www.telegraph.co.uk/.

Morgan, Gareth. *Images of Organization*. 2nd ed. Thousand Oaks, Calif: Sage, 1997.

Morozov, Evgeny. "An Army of Ones and Zeroes." *Slate.com*, August 14, 2008. http://www.slate.com/id/2197514/.

———. "Testimony to the US Commission on Security and Cooperation in Europe," Washington D.C., October 22, 2009. http://csce.gov/.

———. *The Net Delusion: The Dark Side of Internet Freedom*. PublicAffairs, 2011.

Nest, Michael. *Coltan*. Polity, 2011.

Norman, Helmi, and Jillian C. York. "West Censoring East: The Use of Western Technologies by Middle East Censors, 2010–2011." *OpenNet Initiative*, 2010. http://opennet .net/.

O'Connor, Mary Catherine. "Google Maps Finally Adds Bike Routes." *Wired*, March 10, 2010. http://www.wired.com/.

Ong, Walter J. *Orality and Literacy: The Technologizing of the Word.* Methuen, 1982.

Pasquinelli, Matteo. *Animal Spirits: A Bestiary of the Commons.* Rotterdam: NAi Publishers, 2009.

Putnam, Robert D. *Bowling Alone: The Collapse and Revival of American Community.* New York: Simon & Schuster, 2000.

Rancière, Jacques. *Disagreement: Politics and Philosophy.* Minneapolis: University of Minnesota Press, 1999.

———. *On the Shores of Politics: Franco-Brazilian Symposium on Power: Selected Papers.* Translated by Liz Heron. Verso, 1995.

"RBN—Extortion and Denial of Service (DDOS) Attacks." *Russian Business Network,* February 2008. http://rbnexploit.blogspot.com/2008/02/rbn-extortion-and-denial -of-service.html.

Readings, Bill. *Introducing Lyotard: Art and Politics.* Routledge, 1991.

Rheingold, Howard. *Smart Mobs: The Next Social Revolution.* Basic Books, 2003.

Rhoads, Christopher, and Loretta Chao. "Iran's Web Spying Aided by Western Technology." *The Wall Street Journal,* June 22, 2009. http://online.wsj.com/.

Richburg, Keith B. "Google Compromise Pays Off with Renewal of License in China." *The Washington Post,* July 10, 2010. http://www.washingtonpost.com/.

Rivers, Theodore John. "An Introduction to the Metaphysics of Technology." *Technology in Society* 27, no. 4 (November 2005): 551–74.

———. *Contra Technologiam.* University Press of America, 1993.

Robb, John. "Open Source Warfare: Cyberwar." *Global Guerillas,* August 15, 2008. http:// globalguerrillas.typepad.com/.

Roberts, Joseph W. *How the Internet Is Changing the Practice of Politics in the Middle East.* Edwin Mellen Press, 2009.

Rodgers, Jayne. "Doreen Massey." *Information, Communication & Society* 7, no. 2 (January 2001): 273–91.

Rodham Clinton, Hillary. "Remarks on Internet Freedom". U.S. Department of State, January 21, 2010. http://www.state.gov/secretary/rm/2010/01/135519.htm.

Rosen, Jay. "The People Formerly Known as the Audience." *PressThink,* June 27, 2006. http://journalism.nyu.edu/pubzone/weblogs/pressthink/.

Rossiter, Ned. *Organized Networks: Media Theory, Creative Labour, New Institutions.* Rotterdam: NAi Publishers, 2006.

Sardar, Ziauddin, Ashis Nandy, and Merryl Wyn Davies. *Barbaric Others: A Manifesto on Western Racism.* Pluto Press, 1993.

Sassi, Sinikka. "Cultural Differentiation or Social Segregation? Four Approaches to the Digital Divide." *New Media & Society* 7, no. 5 (October 1, 2005): 684–700.

Scholz, Trebor. "What the MySpace Generation Should Know about Working for Free." *Collectivate.net,* April 3, 2007. http://www.collectivate.net/.

Schutz, Alfred. *The Phenomenology of the Social World.* Northwestern University Studies in Phenomenology and Existential Philosophy. Evanston, Ill.: Northwestern University Press, 1967.

Scoble, Robert, and Shel Israel. *Naked Conversations: How Blogs Are Changing the Way Businesses Talk with Customers.* Hoboken, N.J.: John Wiley, 2006.

Seabrook, John. "The Price of the Ticket." *The New Yorker*, August 10, 2009. http://www.newyorker.com/.

Searls, Doc. "Power Re-Origination." *Doc Searls Weblog*, n.d. http://doc-weblogs.com/2006/06/28#powerReorigination.

Segaran, Toby. *Programming Collective Intelligence: Building Smart Web 2.0 Applications*. O'Reilly, 2007.

Serres, Michel. *The Parasite*. Minneapolis: University of Minnesota Press, 2007.

Serrie, Jonathan. "Propaganda War Rages Online." *Fox News On the Scene*, January 8, 2009. http://onthescene.blogs.foxnews.com/.

Shapiro, Samantha M. "Revolution, Facebook-Style." *The New York Times*, January 22, 2009. http://www.nytimes.com/.

Shenker, Jack. "Fury over Advert Claiming Egypt Revolution as Vodafone's." *The Guardian*, June 3, 2011. http://www.guardian.co.uk/.

Shirky, Clay. "Folksonomy." *Many 2 Many*, August 25, 2004. http://many.corante.com/archives/2004/08/25/folksonomy.php.

———. *Here Comes Everybody: The Power of Organizing without Organizations*. Penguin, 2009.

———. "Matt Locke on Folksonomies." *Many 2 Many*, March 1, 2005. http://www.corante.com/many/archives/2005/03/01/matt_locke_on_folksonomies.php.

Silverstone, Roger. *Why Study the Media?* 1st ed. Sage, 1999.

Singer, P. W. *Wired for War: The Robotics Revolution and Conflict in the 21st Century*. Penguin, 2009.

Sliva, Amy, V. S. Subrahmanian, Vanina Martinez, and Gerardo I. Simari. "The SOMA Terror Organization Portal (STOP): Social Network and Analytic Tools for the Real-Time Analysis of Terror Groups." In *Social Computing, Behavioral Modeling, and Prediction*, edited by Huan Liu, John Salerno, and Michael J. Young, 9–18. New York: Springer, 2008.

Smith, Aaron. "Mobile Access 2010." *Pew Internet and American Life Project*, July 7, 2010. http://pewinternet.org/.

Smith, Aaron, and Lee Rainie. *The Internet and the 2008 Election*. Washington, D.C.: Pew Internet and American Life Project, June 15, 2008. http://www.pewinternet.org/.

"Social Networks/Blogs Now Account for One in Every Four and a Half Minutes Online." *Nielsen Wire*, June 15, 2010. http://blog.nielsen.com/nielsenwire/.

Solomon, Norman. *War Made Easy: How Presidents and Pundits Keep Spinning Us to Death*. Wiley, 2006.

Spring, Tom. "Quit Facebook Day Was a Success Even as it Flopped." *PCWorld*, June 1, 2010. http://www.pcworld.com/.

Sützl, Wolfgang. "Tragic Extremes: Nietzsche and the Politics of Security." *cTheory*, 2007. http://www.ctheory.net/articles.aspx?id=582.

Tangney, John, and Judith M. Lytle. "Preface." In *Social Computing, Behavioral Modeling, and Prediction*, edited by Huan Liu, John Salerno, and Michael J. Young, v–viii. New York: Springer, 2008.

Tapscott, Don, and Anthony D. Williams. *Wikinomics: How Mass Collaboration Changes Everything*. Portfolio Trade, 2010.

Tehrani, Hamid. "Iranian Officials 'Crowd-Source' Protester Identities." *Global Voices,* June 27, 2009. http://globalvoicesonline.org/.

Terranova, Tiziana. *Network Culture: Politics for the Information Age.* London: Pluto Press, 2004.

Thierer, Adam, and Grant Eskelsen. *Media Metrics: The True State of the Modern Media Marketplace Version 1.0.* Washington, D.C.: The Progress and Freedom Foundation, 2008. http://www.pff.org/mediametrics/.

Tocqueville, Alexis de. *Democracy in America.* New York: Library of America, 2004.

Tryhorn, Chris. "Nice Talking to You . . . Mobile Phone Use Passes Milestone." *The Guardian,* March 3, 2009. http://www.guardian.co.uk/.

"Valley Boys." *Bloomberg Businessweek,* August 14, 2006. http://www.businessweek .com/.

Vandenberghe, F. E. E. "Reconstructing Humants: A Humanist Critique of Actant-Network Theory." *Theory, Culture & Society* 19, no. 5–6 (2002): 51.

Virno, Paolo. *A Grammar of the Multitude: For an Analysis of Contemporary Forms of Life.* Translated by Isabella Bertoletti, James Cascaito, and Andrea Casson. Cambridge, Mass.: Semiotext(e), 2004.

Walther, Joseph B. "Computer-Mediated Communication." *Communication Research* 23, no. 1 (February 1, 1996): 3–43.

Wark, McKenzie. *Gamer Theory.* Cambridge, Mass.: Harvard University Press, 2007.

Warren, Christina. "Quit Facebook Day Falls Flat." *Mashable,* June 1, 2010. http:// mashable.com/2010/06/01/facebook-quit-results/.

Weis, Allan, and Valerie Andrews. *The Business of Changing Lives.* Greenleaf Book (Distributor), 2009.

Wellman, Barry. "Little Boxes, Glocalization, and Networked Individualism." In *Digital Cities II. Computational and Sociological Approaches: Second Kyoto Workshop on Digital Cities, Kyoto, Japan, October 18–20, 2001. Revised Papers (Lecture Notes in Computer Science),* edited by Makoto Tanabe, Peter van den Besselaar, and Toru Ishida, 11–25. Springer, 2002.

———. "The Network Community: An Introduction." In *Networks in the Global Village: Life in Contemporary Communities,* edited by Barry Wellman, 1–48. Boulder, Colo.: Westview, 1998.

———, ed. *Networks in the Global Village: Life in Contemporary Communities.* Boulder, Colo.: Westview, 1998.

"What This Essay Is About". P2P Fundation, January 18, 2006. http://p2pfoundation .net/What_this_essay_is_about.

White, Micah M. "Facebook Suicide." *Adbusters,* June 4, 2008. http://www.adbusters .org/magazine/77/facebook_suicide.html.

"Wikipedia: Neutral Point of View." *The Wikipedia,* n.d. http://en.wikipedia.org/wiki/ Wikipedia:NPOV.

Wilkinson, Alec. "Non-Lethal Force." *The New Yorker,* June 2, 2008. http://www.new yorker.com/.

Winner, Langdon. *Autonomous Technology: Technics-Out-of-Control as a Theme in Political Thought.* Cambridge, Mass.: MIT Press, 1978.

Wise, J. Macgregor. *Exploring Technology and Social Space*. Sage, 1997.

Wong, Queenie. "Social Media Play Big Role in Riot Probe." *The Seattle Times*, June 16, 2011. http://seattletimes.nwsource.com/.

Wu, Tim. *The Master Switch: The Rise and Fall of Information Empires*. Knopf, 2010.

Wyatt, Sally, Graham Thomas, and Tiziana Terranova. "They Came, They Surfed, They Went Back to the Beach: Conceptualizing Use and Non-Use of the Internet." In *Virtual Society? Technology, Cyberbole, Reality*, edited by Steve Woolgar, 23–40. Oxford: Oxford University Press, 2002.

York, Jillian C. "Syria's Twitter Spambots." *The Guardian*, April 21, 2011. http://www.guardian.co.uk/.

———. "This Week in Internet Censorship." *Electronic Frontier Foundation*, June 15, 2011. https://www.eff.org/.

Youth and Downloading Behavior. Business Software Alliance/Harris Interactive, 2007. http://www.bsa.org/.

Yus, Francisco. "The Linguistic-Cognitive Essence of Virtual Community." *Iberica* 9 (2005): 79–102.

Zuckerman, Ethan. "Does the Number Have a Lesson for Human Rights Activists?" *WorldChanging.com*, May 3, 2007. http://www.worldchanging.com/archives/006626.html

INDEX

nodes: properties, 42
nodocentrism, 9–13, 16, 45–46, 48, 83, 87, 101, 133, 156–57
noise and communication, 16–17

OBAMM. *See* Online Behavioral Analysis and Modeling Methadology (OBAMM)
Online Behavioral Analysis and Modeling Methadology (OBAMM), 103
open content, 123
OpenNet Initiative, 109
ownership and control of networks, 19

Pandith, Farah, 140
paradoxes of network logic, 90
paralogies, 90–91, 99
paranodes, 115, 153–54, 159
parasitology, 90
participation in networks, 4, 25–27
participatory media, 30–33
participatory war, 116–18
Pasquinelli, Matteo, 123, 129
peer-to-peer (P2P) sharing: as alternative, 126; cyberpiracy, 131–33; decentralized structure, 126–29; limitations, 128–29; new model of production, 123; "new socialism," 129–31; nodocentrism, 133
Plato, 58
playbor, 8, 25
political considerations: blocking access to networks, 102; Internet freedom, 134–36; networks and political uprisings, 107; tools for activism, 104
political uprisings and networks, 107
preferential attachment process, 4
privacy issues: activists, 138; e-mail concerns, 7; Internet freedom, 136; sacrifice of privacy, 5, 20, 26, 75; settings, 61, 158; surveillance and privacy, 109–10
privatization: aspect of commodification, 22–23; social spaces, 3

protesters and political dissent, 111
proximity and distance, 97–98, 101
PSYOPs, 109
P2P sharing. *See* peer-to-peer (P2P) sharing
public and private media, 30
publics and masses, 68–70

Quit Facebook Day, xi–xii

radio frequency identification (RFID) blockers, 154
Rancière, Jacques, 112, 154
Rashid, Esraa Abdel Fattah Ahmed, 108
realism, 64
RFID blockers. *See* radio frequency identification (RFID) blockers
Rivers, Theodore, 63, 70, 89
Robb, John, 117
Rosen, Jay, 31
Ross, Alec, 140
Rossiter, Ned, 81
Rumsfeld, Donald, 113

SAFE. *See* SOMA Adversarial Forecast Engine (SAFE)
SANE. See SOMA Analyst NEtwork (SANE)
scale-free networks, 41, 156
Scholz, Trebor, 40
Schutz, Alfred, 61–62
Searls, Doc, 133
security: networked, 113–16; personal, 21, 71, 75, 138; surveillance, 11, 103, 106
SEE. *See* SOMA Extraction Engine (SEE)
Shapiro, Samantha M., 107
Shirky, Clay, 52
Silverstone, Roger, 97
simultaneity and network mediation, 61–62
slow food movement, 159
social allegories, 46–47
social capital, 44
social computing, 39–40

(continued from page ii)

ULISES ALI MEJIAS is assistant professor of communication studies at the State University of New York, College at Oswego.